TUNNELS AND UNDERGROUND CITIES: ENGINEERING AND INNOVATION MEET ARCHAEOLOGY, ARCHITECTURE AND ART

PROCEEDINGS OF THE WTC2019 ITA-AITES WORLD TUNNEL CONGRESS, NAPLES, ITALY, 3-9 MAY, 2019

Tunnels and Underground Cities: Engineering and Innovation meet Archaeology, Architecture and Art

Volume 1: Archeology, Architecture and Art in underground construction

Editors

Daniele Peila

Politecnico di Torino, Italy

Giulia Viggiani

University of Cambridge, UK
Università di Roma "Tor Vergata", Italy

Tarcisio Celestino

University of Sao Paulo, Brasil

CRC Press is an imprint of the
Taylor & Francis Group, an **informa** business

A BALKEMA BOOK

Cover illustration:

View of Naples gulf

CRC Press/Balkema is an imprint of the Taylor & Francis Group, an informa business

Typeset by Integra Software Services Pvt. Ltd., Pondicherry, India

Published by: CRC Press/Balkema
Schipholweg 107C, 2316XC Leiden, The Netherlands
e-mail: Pub.NL@taylorandfrancis.com
www.crcpress.com – www.taylorandfrancis.com

ISBN: 978-0-367-46574-2 (Hbk)
ISBN: 978-1-003-02967-0 (eBook)

Tunnels and Underground Cities: Engineering and Innovation meet Archaeology, Architecture and Art, Volume 1: Archeology, Architecture and Art in underground construction – Peila, Viggiani & Celestino (Eds)
© 2020 Taylor & Francis Group, London, ISBN 978-0-367-46574-2

Table of contents

Preface vii

Acknowledgements ix

WTC 2019 Congress Organization xi

Archeology, Architecture and Art in underground construction 1

Reflecting the art, culture and history into the interior design of underground metro stations, case study: Dnipro metro extension line 3
B. Avanoğlu-Çetin & Ö. Öztürk

The Art Stations 12
F. Brenci

The outlets of the Albano and Fucine Lakes and their influence from Roman times up to the birth of modern tunnel engineering 22
C. Callari

Urban tunnelling under archaeological findings in Naples (Italy) with ground freezing and grouting techniques 32
F. Cavuoto, V. Manassero, G. Russo & A. Corbo

Integration of archaeology in architectural design of Milan Metro connection M2–M4 in St. Ambrogio Station 42
M.N. Colombo, A. Bortolussi & E. Noce

Tunnel warfare in World War I: The underground battlefield tunnels of Vimy Ridge, France 52
M. Diederichs & D.J. Hutchinson

Archaeology in underground construction: The experience acquired during construction of Italian high-speed railway lines 62
F. Frandi

The architecture of underground dwellings in Iran 72
S. Hashemi

Spiritual life and life after death in the undergrounds of ancient Iran 80
S. Hashemi

The Albinian way of design at the Milan Metro 90
Y. Kutkan-Öztürk

Mobilizing cultural resources: The functional role of heritage in metro projects 98
M. Laudato

The archeological evidences of the De Amicis Station in the Milan Metro line 4, Italy 108
G. Lunardi, G. Cassani, M. Gatti & S. Gazzola

Archaeology and tunnelling interaction in the railway project of Catania underpass in Sicily, Italy 118
E. Manfredi, F. Iannotta, F. Romano & S. Vanfiori

A workflow process for tunnels maintenance. The case of the Construction Method developed for Rhaetian Railways (UNESCO World Heritage Site) 127
F. Modetta, A. Arigoni, S. Saviani, K. Grossauer & M. Hohermuth

The First World War military tunnels of the Italian-Austrian front 137
S. Pedemonte & E.M. Pizzarotti

Metro Thessaloniki – intersecting microtunnels to support archeological findings at Sintrivani station 147
D. Rizos, G. Vassilakopoulou, P. Foufas & G. Anagnostou

Rome and its stratification 157
G. Romagnoli

Line C in Rome: San Giovanni, the first archaeological station 167
E. Romani, M. D'Angelo & V. Foti

The archaeological findings are changing Amba-Aradam station design in Rome Metro Line C 178
E. Romani, M. D'Angelo & R. Sorge

Underground car park in the ancient "Morelli" cavern in Naples 188
F. Rossano, A. Bellone & M.A. Piangatelli

Moncenisio, from Myth to history TELT – Tunnel Euralpin Lyon Turin and the collection of historic engravings on the Frejus tunnel 199
M. Virano, G. Dati, M. Ricci & G. Avataneo

Interdisciplinary research in geotechnical engineering and geoarchaeology – a London case study 209
F.K. Vonstad, P. Ferreira & D. Sully

Author index

Tunnels and Underground Cities: Engineering and Innovation meet Archaeology, Architecture and Art, Volume 1: Archeology, Architecture and Art in underground construction – Peila, Viggiani & Celestino (Eds)
© 2020 Taylor & Francis Group, London, ISBN 978-0-367-46574-2

Preface

The World Tunnel Congress 2019 and the 45th General Assembly of the International Tunnelling and Underground Space Association (ITA), will be held in Naples, Italy next May.

The Italian Tunnelling Society is honored and proud to host this outstanding event of the international tunnelling community.

Hopefully hundreds of experts, engineers, architects, geologists, consultants, contractors, designers, clients, suppliers, manufacturers will come and meet together in Naples to share knowledge, experience and business, enjoying the atmosphere of culture, technology and good living of this historic city, full of marvelous natural, artistic and historical treasures together with new innovative and high standard underground infrastructures.

The city of Naples was the inspirational venue of this conference, starting from the title Tunnels and Underground cities: engineering and innovation meet Archaeology, Architecture and Art.

Naples is a cradle of underground works with an extended network of Greek and Roman tunnels and underground cavities dated to the fourth century BC, but also a vibrant and innovative city boasting a modern and efficient underground transit system, whose stations represent one of the most interesting Italian experiments on the permanent insertion of contemporary artwork in the urban context.

All this has inspired and deeply enriched the scientific contributions received from authors coming from over 50 different countries.

We have entrusted the WTC2019 proceedings to an editorial board of 3 professors skilled in the field of tunneling, engineering, geotechnics and geomechanics of soil and rocks, well known at international level. They have relied on a Scientific Committee made up of 11 Topic Coordinators and more than 100 national and international experts: they have reviewed more than 1.000 abstracts and 750 papers, to end up with the publication of about 670 papers, inserted in this WTC2019 proceedings.

According to the Scientific Board statement we believe these proceedings can be a valuable text in the development of the art and science of engineering and construction of underground works even with reference to the subject matters "Archaeology, Architecture and Art" proposed by the innovative title of the congress, which have "contaminated" and enriched many proceedings' papers.

Andrea Pigorini
SIG President

Renato Casale
Chairman of the Organizing Committee WTC2019

Tunnels and Underground Cities: Engineering and Innovation meet Archaeology, Architecture and Art, Volume 1: Archeology, Architecture and Art in underground construction – Peila, Viggiani & Celestino (Eds)
© 2020 Taylor & Francis Group, London, ISBN 978-0-367-46574-2

Acknowledgements

REVIEWERS

The Editors wish to express their gratitude to the eleven Topic Coordinators: Lorenzo Brino, Giovanna Cassani, Alessandra De Cesaris, Pietro Jarre, Donato Ludovici, Vittorio Manassero, Matthias Neuenschwander, Moreno Pescara, Enrico Maria Pizzarotti, Tatiana Rotonda, Alessandra Sciotti and all the Scientific Committee members for their effort and valuable time.

SPONSORS

The WTC2019 Organizing Committee and the Editors wish to express their gratitude to the congress sponsors for their help and support.

Tunnels and Underground Cities: Engineering and Innovation meet Archaeology, Architecture and Art, Volume 1: Archeology, Architecture and Art in underground construction – Peila, Viggiani & Celestino (Eds)
© 2020 Taylor & Francis Group, London, ISBN 978-0-367-46574-2

WTC 2019 Congress Organization

HONORARY ADVISORY PANEL

Pietro Lunardi, President WTC2001 Milan
Sebastiano Pelizza, ITA Past President 1996-1998
Bruno Pigorini, President WTC1986 Florence

INTERNATIONAL STEERING COMMITTEE

Giuseppe Lunardi, Italy (Coordinator)
Tarcisio Celestino, Brazil (ITA President)
Soren Eskesen, Denmark (ITA Past President)
Alexandre Gomes, Chile (ITA Vice President)
Ruth Haug, Norway (ITA Vice President)
Eric Leca, France (ITA Vice President)
Jenny Yan, China (ITA Vice President)
Felix Amberg, Switzerland
Lars Barbendererder, Germany
Arnold Dix, Australia
Randall Essex, USA
Pekka Nieminen, Finland
Dr Ooi Teik Aun, Malaysia
Chung-Sik Yoo, Korea
Davorin Kolic, Croatia
Olivier Vion, France
Miguel Fernandez-Bollo, Spain (AETOS)
Yann Leblais, France (AFTES)
Johan Mignon, Belgium (ABTUS)
Xavier Roulet, Switzerland (STS)
Joao Bilé Serra, Portugal (CPT)
Martin Bosshard, Switzerland
Luzi R. Gruber, Switzerland

EXECUTIVE COMMITTEE

Renato Casale (Organizing Committee President)
Andrea Pigorini, (SIG President)
Olivier Vion (ITA Executive Director)
Francesco Bellone
Anna Bortolussi
Massimiliano Bringiotti
Ignazio Carbone
Antonello De Risi
Anna Forciniti
Giuseppe M. Gaspari

Giuseppe Lunardi
Daniele Martinelli
Giuseppe Molisso
Daniele Peila
Enrico Maria Pizzarotti
Marco Ranieri

ORGANIZING COMMITTEE

Enrico Luigi Arini
Joseph Attias
Margherita Bellone
Claude Berenguier
Filippo Bonasso
Massimo Concilia
Matteo d'Aloja
Enrico Dal Negro
Gianluca Dati
Giovanni Giacomin
Aniello A. Giamundo
Mario Giovanni Lampiano
Pompeo Levanto
Mario Lodigiani
Maurizio Marchionni
Davide Mardegan
Paolo Mazzalai
Gian Luca Menchini
Alessandro Micheli
Cesare Salvadori
Stelvio Santarelli
Andrea Sciotti
Alberto Selleri
Patrizio Torta
Daniele Vanni

SCIENTIFIC COMMITTEE

Daniele Peila, Italy (Chair)
Giulia Viggiani, Italy (Chair)
Tarcisio Celestino, Brazil (Chair)
Lorenzo Brino, Italy
Giovanna Cassani, Italy
Alessandra De Cesaris, Italy
Pietro Jarre, Italy
Donato Ludovici, Italy
Vittorio Manassero, Italy
Matthias Neuenschwander, Switzerland
Moreno Pescara, Italy
Enrico Maria Pizzarotti, Italy
Tatiana Rotonda, Italy
Alessandra Sciotti, Italy
Han Admiraal, The Netherlands
Luisa Alfieri, Italy
Georgios Anagnostou, Switzerland
Andre Assis, Brazil
Stefano Aversa, Italy
Jonathan Baber, USA
Monica Barbero, Italy
Carlo Bardani, Italy
Mikhail Belenkiy, Russia
Paolo Berry, Italy
Adam Bezuijen, Belgium
Nhu Bilgin, Turkey
Emilio Bilotta, Italy
Nikolai Bobylev, United Kingdom
Romano Borchiellini, Italy
Martin Bosshard, Switzerland
Francesca Bozzano, Italy
Wout Broere, The Netherlands
Domenico Calcaterra, Italy
Carlo Callari, Italy

Luigi Callisto, Italy
Elena Chiriotti, France
Massimo Coli, Italy
Franco Cucchi, Italy
Paolo Cucino, Italy
Stefano De Caro, Italy
Bart De Pauw, Belgium
Michel Deffayet, France
Nicola Della Valle, Spain
Riccardo Dell'Osso, Italy
Claudio Di Prisco, Italy
Arnold Dix, Australia
Amanda Elioff, USA
Carolina Ercolani, Italy
Adriano Fava, Italy
Sebastiano Foti, Italy
Piergiuseppe Froldi, Italy
Brian Fulcher, USA
Stefano Fuoco, Italy
Robert Galler, Austria
Piergiorgio Grasso, Italy
Alessandro Graziani, Italy
Lamberto Griffini, Italy
Eivind Grov, Norway
Zhu Hehua, China
Georgios Kalamaras, Italy
Jurij Karlovsek, Australia
Donald Lamont, United Kingdom
Albino Lembo Fazio, Italy
Roland Leucker, Germany
Stefano Lo Russo, Italy
Sindre Log, USA
Robert Mair, United Kingdom
Alessandro Mandolini, Italy
Francesco Marchese, Italy
Paul Marinos, Greece
Daniele Martinelli, Italy
Antonello Martino, Italy
Alberto Meda, Italy
Davide Merlini, Switzerland
Alessandro Micheli, Italy
Salvatore Miliziano, Italy
Mike Mooney, USA
Alberto Morino, Italy
Martin Muncke, Austria
Nasri Munfah, USA
Bjørn Nilsen, Norway
Fabio Oliva, Italy
Anna Osello, Italy
Alessandro Pagliaroli, Italy
Mario Patrucco, Italy
Francesco Peduto, Italy
Giorgio Piaggio, Chile
Giovanni Plizzari, Italy
Sebastiano Rampello, Italy
Jan Rohed, Norway
Jamal Rostami, USA
Henry Russell, USA
Giampiero Russo, Italy
Gabriele Scarascia Mugnozza, Italy
Claudio Scavia, Italy
Ken Schotte, Belgium
Gerard Seingre, Switzerland
Alberto Selleri, Italy
Anna Siemińska Lewandowska, Poland
Achille Sorlini, Italy
Ray Sterling, USA
Markus Thewes, Germany
Jean-François Thimus, Belgium
Paolo Tommasi, Italy
Daniele Vanni, Italy
Francesco Venza, Italy
Luca Verrucci, Italy
Mario Virano, Italy
Harald Wagner, Thailand
Bai Yun, China
Jian Zhao, Australia
Raffaele Zurlo, Italy

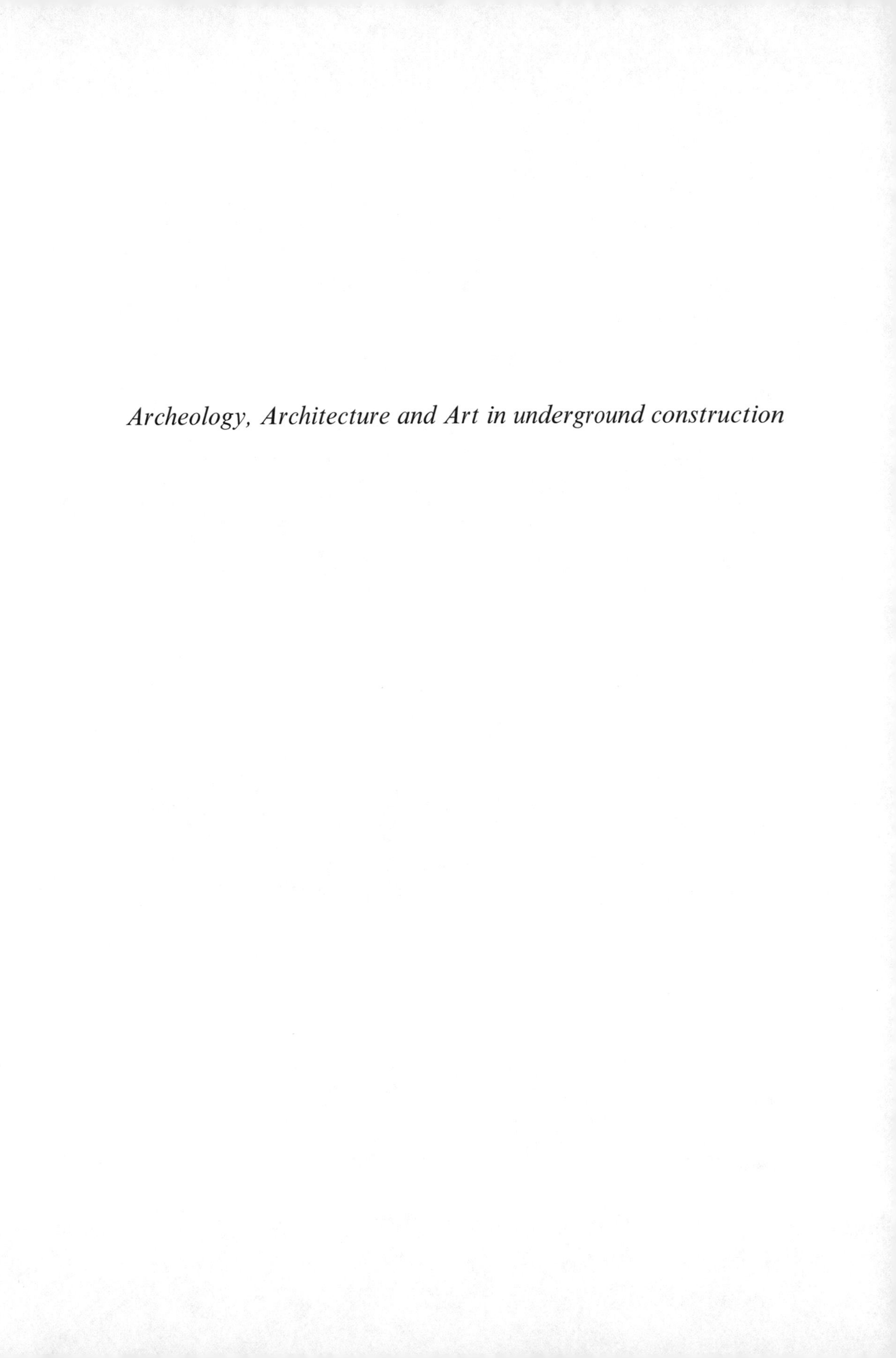

Archeology, Architecture and Art in underground construction

Tunnels and Underground Cities: Engineering and Innovation meet Archaeology, Architecture and Art, Volume 1: Archeology, Architecture and Art in underground construction – Peila, Viggiani & Celestino (Eds)
 ISBN 978-0-367-46574-2

Reflecting the art, culture and history into the interior design of underground metro stations, case study: Dnipro metro extension line

B. Avanoğlu-Çetin & Ö. Öztürk
Yüksel Proje Uluslararası A.Ş., Ankara, Turkey

ABSTRACT: Three new stations designed for the Dnipro subway extension line, Ukraine; Theatre, Central and the Museum Stations. The main intention was to create unique interior designs for each station. By conducting a detailed study of the local historical narratives and cultural themes of the city, interior design concepts developed so that a station building could be more than a portal for transportation needs. This paper will present the design approach and considerations for the stations and their inspirational stories. Moreover, readers can access the designs including panoramic views by using their smartphones as a new architectural presentation platform.

1 INTRODUCTION

The city of Dnipro, located on the banks of Dnieper River in central Ukraine. It has an existing subway line built in 1995 and consists of six stations. In 2016 with an extension to the existing line, three more stations designed for the city, Theatre Station, Central Station and Museum Station. The project officially started on July 2016 and expected to be finished in 2021 ('Dnipro metro extension contract signed - Railway Gazette', 2016).

As it is possible to see from the names of the stations, they are situated in important urban nodes. Stations locations are on the centre of the city where one can see an important volume of passengers on the move not only for their daily routine paths but also for different cultural and leisure activities. Extension line follows the Dmytra Yavornytskoho Avenue under the ground. Each of the stations have similar layouts in terms of ticket hall/concourse and platform areas. Yet they all inspired by different ideas due to their different contexts.

The first station of the extension line is the Theatre (Teatralnaya) Station located very close to the Dniprovsky Academic Drama & Comedy Theatre and other various theatres in the city. Next station is the Central (Centralnaya) Station, which is located at the centre of the city. Subsequently, the last station on the extension line is Historical Museum Station.

This paper will present the main interior design process of the stations and show how historical narratives and cultural themes of the city influenced the design ideas. Firstly, the main inspirations and the framework of the design will be introduced. Afterwards, every station will be presented individually with its design decisions and themes.

2 INSPIRATIONS

2.1 *Ukrainian flag and coat of arms*

Before starting with the design process of the stations, it is important to define the main frame of the concepts. The history and the geography of Ukraine considered in the process. Firstly, the national flag of the country and its meaning are pointed out as some of the basic notions about the design. In the simplest explanation, the flag of Ukraine is two equally sized parallel bands of blue and yellow. Yellow colour represents wheat fields that refer to prosperity and bountifullness, and the colour of blue signifies peace and blue skies. Its dynamic colours and peaceful meanings were one of our starting points for the design process.

Like the national flag, Ukrainian Coat of Arms features the same colours, a blue shield with a gold pitchfork. In the beginning, it was the seal-trident of Volodymyr the Great and afterwards used by different rulers throughout time. Currently, it is possible to see it around various buildings from different periods of Ukraine. Both the colour scheme of the national flag and abstractions of the coat of arms integrated to the general design approaches with their respective themes.

Figure 1. Ukrainian coat of arms and its stylized versions for each station. From left to right; the official coat of arms, stylized version of theatre station, central station and historical museum station (author).

2.2 *Dnieper river and agriculture*

In addition to the national flag and the coat of arms, Dnieper River and agriculture are major factors that not only shapes the geography but also its cultural background.

Agriculture is a significant part of Ukraine's economy and as mentioned above yellow part of the national flag is depicted as the fertile wheat fields of the country. Not only in this specific design case study but also during the conducted research for the design process it is possible to see that the fertile lands and its agricultural significance shaped the culture and its implications on the local arts. Agriculture and its reflections on the local art of the society pointed out the rich and vast culture of the area.

However, it would be an erroneous act to think about Ukraine without the Dnieper River. Dnieper River and its basin shaped the geography and nourished the population on every aspect. It is possible to trace the Dnieper at the national anthem of Ukraine, Grandfather Dnipro, in Gogol's stories (Gogol, Pevear and Volokhonsky, 1998) or in Aivazovsky's

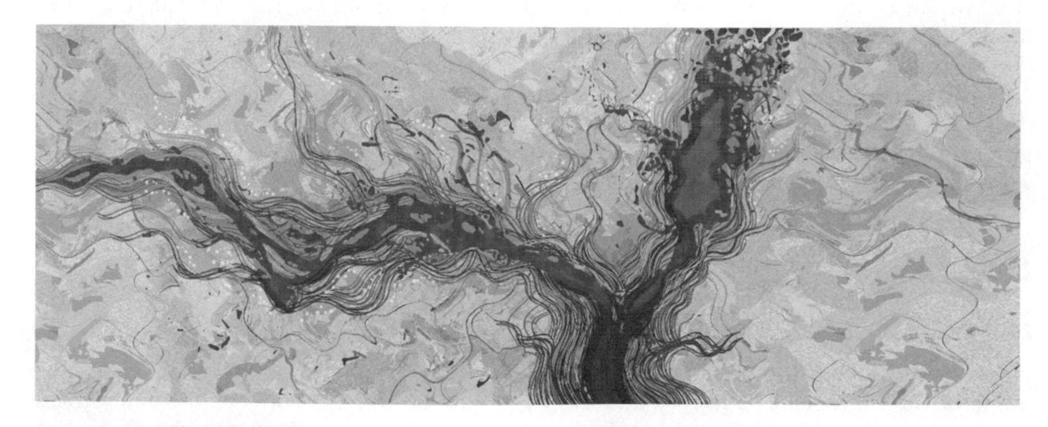

Figure 2. Dnipro river artwork design for central station (author).

paintings. As a result, the river itself and its meaning was a major element in the process of the design and incorporated into the project.

3 STATIONS

Three new stations are going to be constructed with the extension of the existing subway line of Dnipro. As mentioned before, the extension line follows the existing urban paths connecting essential parts of the city. Name of the stations are; the Theatre (Teatralnaya) Station, the Central (Centralnaya) Station and the Historical Museum Station. A general approach is applied to the technical layout of the stations. Pedestrians enter the station from a centralized entry point where they can perceive the concourse and its ongoing activities. Centralized entry point enables to establish a linear approach to the pedestrian movement. Security, waiting areas and ticketing machines are placed to the sides to avoid obstruction to passenger flow. In addition, an agent-based pedestrian simulation was set up to verify the selected design approach. The concourse is separated into two areas with the ticket tolls, as paid and unpaid areas. Following the concourse area, the escalator tunnel leads the passengers to the platform level.

Different themes are selected for each station so that an underground structure used for mass transportation could tell a story about its context to the passengers. These are; Ukrainian Avant-garde for the Theatre Station, Petrykivka Paintings and Vyshyvanka Patterns for the Central Station and the Ancient Cultures for the Museum Station.

3.1 *Ukrainian avant-garde and the theatre station*

During the design process, one of the key aspects of the design was to connect some wide extents and themes to its context. On behalf of the Theatre Station, Ukrainian Theatre and the 20th Ukrainian Avant-garde is selected because of its dramatic and significant narrative. Starting with the Soviet Revolution of 1917 this inspirational art movement lived its heyday between 1920-1930ies; until it clashed with the State and state-sponsored Socialist Realism, it included various artists including Kazimir Malevich, David Burliuk, Vladimir Tatlin, and Aleksandra Ekster who were connected to Ukraine by birth, education or identity. However, due to tragic events and political repression, this era is also described as the "Executed Renaissance"(Hryn, 2004). It is possible to see pioneering attempts to develop the Ukrainian literature and fine arts with not only passionate contemporary attempts but also adaptations of various Shakespeare, Sinclair, Victor Hugo and Friedrich Schiller plays(Onyshkevych, 1999). As a result, with a theatrical setting certain names underscored for this design; Berezil Theatre, Les Kurbas, Anatol Petrytsky, Alexandra Ekster and Vadym Meller.

Les Kurbas (1887-1937) and Vadym Meller's (1884-1962) stage designs at the Berezil (Spring/New Beginning) Theatre play a crucial role in the design of the station. Their inspiring "movement", which was reflected the stage and costume design, inspired new pioneering techniques that helped to establish a new approach(Mudrak, 1981). Additionally, by staging the translated classic playwrights and incorporating new Ukrainian plays, Berezil helped to establish the notion of a national theatre in its context. With Anatol Petrytsky(1895-1964) and Alexandra Ekster's (1882-1949) stage designs incorporated into the theatre, it is possible to see a richness and movement at the stage(Horbachov, 2010). However, with the political shifts artists who support the movement accused of being nationalist, rationalist, formalist, and counterrevolutionary. Most of the works and stages were destroyed in the 1930s (Hryn, 2004). Their remaining sketches, paintings and drawings give us a glimpse of what they tried to establish for their artistic view. By looking at this dramatic history of theatre in Ukraine, with the introduction of dynamic and inspiring stage and costume designs set the tone for the whole frame with given question what if this "theatre" is their new stage and every passenger step up to their stage.

As a result, the whole idea is around the fact that the dynamic movement of a subway station has a great potential to experience the settings of the Berezil Theatre. By comparing the contemporary stage and lighting design, the main intention of the concept is to set up the "stage" in the platform area and design the concourse are as a foyer. It is decided that all the artwork placed on the walls should be reproductions of the original drawings of the selected artists in order to pay homage. Only the floor textures are adopted from the original artworks.

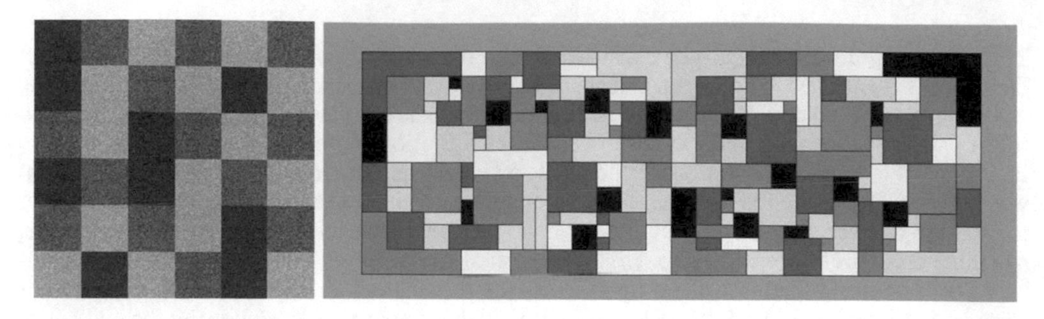

Figure 3. Designed floor textures for theatre station (author).

Artworks are applied to the selected areas to create the "stage" and the "foyer" effect with different techniques. Paintings are placed at the concourse level of the station, character and costume sketches are placed at the platform level. An abstraction of Ukrainian Coat of Arms in the style of Avant-garde is placed at the stairs leading to the platform level. To emphasize the stage effect on the platform area, a special lighting system designed as well as painting the ceiling to black colour so that passengers can experience the sense of being at a stage. Thus, the theatrical narrative of their own national theatre welcomes passengers in the Theatre station.

Figure 4. Artistic renderings of theatre station (author).

3.2 *Petrykivka paintings and vyshyvanka patterns for the central station*

The second station on the extension line is the Central Station that is located at the Tsentralna Street. This location is regarded as a busy local trading node with various transportation options. Due to the daily activities at the vicinity, it is decided that the theme of the station should contain folkloric themes.

Figure 5. Designed artwork for Central Station platform ceiling inspired from Petrykivka paintings (author).

Ukraine has very deep traditions and cultural activities. Among its rich and colourful arts and traditions, Petrykivka Paintings and the Vyshyvanka Embroideries accentuate in Dnipro oblast. Therefore, these traditions and their artistic concerns are applied to the concept design of the project.

Petrykivka painting or “Petrykivsky” is traditional Ukrainian decorative painting, which originated from Dnipropetrovsk oblast of Ukraine, Petrykivka village. In 2013 UNESCO declared “Petrykivka decorative painting as a phenomenon of the Ukrainian ornamental folk art” as Representative List of the Intangible Cultural Heritage of Humanity (2015). As a result, a small town of the Dnipropetrovsk region has become a very important cultural focal point of Dnipro and Ukraine. Originally, these designs were used to decorate traditional stoves, the main wall of the house, and the frames of windows and doors. Through time, these ornamentations applied to household items with various techniques and forms in a continuous development. Petrykivsky has diverse allegories for example; the rooster stands for fire and spiritual awakening, while birds represent light, harmony, happiness etc (Snikhovska, 2017).

Another important folkloric art and tradition in the area is Vyshyvanka Embroidery. In contrast to Petrykivka paintings, Vyshyvanka Embroideries find their colours from the local dyes (like leaves, berries, roots, flowers...) which reflects the colour of nature. It is possible to find various techniques and local styles with different patterns and colour pallets. Yet, it is also kept in mind that these ornamentations have their reflections on the religious context too. For example, it is believed that some patterns are for religious rituals and cultural celebrations. Another pattern is to keep the weaver from the evil or the evil eyes. Every aspect of the geometry, colour and its location is important and it is possible to find out a combination of these elements in this ornamentation style (Snikhovska, 2017). As a result, these ornamental folk arts, which has very rich symbolism, gave inspiration to the design process of the station.

Figure 6. Artwork for central station concourse level inspired from vyshyvanka embroideries (author).

Figure 7. Artistic renderings of central station (author).

Consequently, main intention of the station design was to tell the folkloric stories embodied in the forms of patterns of the embroideries and paintings. As a result, the concept of the station was placed around creating artworks from these motives and design a cultural palette for its users. In this context, many digital artworks are designed inspired from Petrykivka and Vyshyvanka arts. After a detailed design process, it was possible to enlarge these digital artworks to its considerable sizes. Due to similar architectural layouts, artworks are placed on the walls of the concourse, however, at the platform level; a whole Petrykivka painting around 100-meter-long is placed to the ceiling. Besides, a stone mosaic inspired from the wheats symbolizing fertility is located at the centre of the platform and an abstracted version of Ukrainian Coat of Arms in the style of a Petrykivka painting is placed at the stairs leading to the platform level. As a result, passengers are welcomed to a station that is committed to the memory of the city and its people.

3.3 *Ancient cultures for the historical museum station*

Ukraine hosted many civilizations in its rich history. Among them, we can see Cucuteni-Trypillian (5200 - 3500 BC), Scythians (700-200), Kievan Rus(882–1054), Cuman- Kipchaks (10 century – 1241) (Magocsi, 2010). Remarkable artworks are exhibited at the historical museums of Dnipro. Motifs, images, ceramics, stonework and sculptures have been an inspiration to us in the design process of the Historical Museum Station at Dnipro. Yet two of the cultures, Cucuteni-Trypillian and Cuman-Kipchak, are selected because of their rich heritage.

Cucuteni-Trypillian is a Neolithic culture flourished in modern day Ukraine, Moldovia and Romania between 5200 - 3500 BC (Koroleva, 2011). Discovered images and sculptures on various objects can be seen at different museums. It is assumed that Cucuteni-Trypillian civilization created these symbols by associating the woman with the holiness of the mother and the fertility of the earth (Kordysh, 1953). Therefore, Cucuteni-Trypillian civilization sculptures and other archeologic discoveries inspired the artworks of the Historical Museum Station ticket hall walls. Hitherto, graphics represented at these areas are specially designed for Dnipro Metro Project.

Cuman-Kipchaks were a confederation settled in Ukraine steppes from 11th to 13th centuries. Evidence suggests that they migrated westward through the İrtysh river, settled in modern southern Russia and Ukraine. Stone was one the main material to reflect the culture of the

Figure 8. Designed artwork for museum station inspired from ancient civilizations (author).

tribes and stone sculptures called "Balbal" or "Stone Baba" and they can be found in large numbers in Ukraine(Altinkaynak, 2008). Currently, Dnipro Museum hosts 70 of them in various heights and figures.

When we look at the history of these sculptures, we understand that these steles show different typologies in different regions, but there are common motifs and characteristics that provide us to understand the Kipchaks Balbals. Mostly the figures are depicted with Old Turkic clothing holding a bowl for different purposes. Male statues depicted with moustaches or beards and female statues with small lips. It is also important to know that Kipchaks stone sculptures are a part of the mother goddess cult of Central Asia and it is possible to see Umay, the goddess of fertility in Turkic mythology, among these sculptures (Belli, 2016).

Cucuteni-Trypillian and Kipchaks sculptures mostly seen in the museums in this area have been a reference to us in the design process of the station. At the concourse level, to have an efficient pedestrian flow, artworks inspired by the Cucuteni-Trypillian and Kipchaks sculptures placed at the periphery of the area. For this reason, artworks at the design were inspired by the stonework sculptures of the Historical Museum and replicas of Kipchak sculptures placed in front of the columns at the platform level. In addition, graphic design version of the wheat fields placed at the centre of the platform floors to symbolize the fertile fields of

Figure 9. Artistic renderings of historical museum station (author).

Ukraine. Also, an abstraction of Ukrainian Coat of Arms was placed at the stairs leading to the platform level to create a comprehensive effect of its rich history depicting with wings to symbolize each civilization.

3.4 *VR experience*

During the design process, it was very important to narrate the design ideas not only to the design teams but also to the people who has no technical background.

Computer technologies now allow architects not only storing or retrieving information but also visualizing their design ideas in ways that are more effective. Through digital models, designers and clients can visualize and experience their buildings in its context.

Lately, it is common that breaking the traditional way of two-dimensional presentations and using new technologies to explain design ideas in a more effective way. One of the possibility is the integration of virtual reality (VR) to architectural presentations. Alas, the required hardware for such presentation still has its difficulties. Instead of giving everyone a VR device we decided to turn his or her personal smartphones into a VR presentation medium. Therefore, a smart but effective presentation is developed for the concept design of the project. Still virtual reality images of the designs are uploaded to a server where people can access with their smartphones. By doing so, it is possible to share the design ideas to everyone. Additionally, viewers can also share the same experience to each other which will help to explain the design ideas to a broader community. To access a sample of the presentation readers can access to the presentation from the QR code below.

Figure 10. QR code to access panoramic views (author).

4 CONCLUSION

Stations are regarded as transitional spaces where people are in constant motion. However, stations should be considered as public spaces and they should be designed with that notion. Transportation structures are used by high volumes of passengers on daily basis and they have a direct effect on the psychology of its users. Therefore, it is quintessential to design these spaces considering their effects on the passengers.

In conclusion, the history of Dnipro has exceptional potential to tell astonishing stories about its context. Therefore, while designing subway stations for the city historical narratives and cultural themes of the city helped us to shape interior designs. Main design theme was shaped around Ukraine's rich history, geographic features and the local arts. In addition, specific themes are selected for each of the stations. These are; Ukrainian Avant-garde for the Theatre Station, Petrykivka Paintings and Vyshyvanka Patterns for the Central Station and the Ancient Cultures for the Museum Station. The main intention of the design was to turn transit structures into places where passengers find pieces of memories that not only connects them to the station but also to each other. As a result, three stations of the extension line for Dnipro subway is designed to build a bridge between its passengers and the memory of the space.

REFERENCES

Altinkaynak, E. 2008. The Balbals Of Deşt-I Kıpçak. *Turkish Studies International Periodical For the Languages, Literature and History of Turkish or Turkic*, 3(4).

Belli, O. 2003. Stone Statues and Balbals in Turkic World. *Turkish Academy of Sciences Journal of Archaeology*, (6): 85-116.

Railway Gazette. 2018. *Dnipro metro extension contract signed.* [online] Available at: http://www.railway gazette.com/news/single-view/view/dnipro-metro-extension-contract-signed.html [Accessed 15 Sep. 2018].

Gogol, N. 1999. The Terrible Vengeance. *The Collected Tales of Nikolai Gogol.* New York, NY: Vintage Books, 64-105.

Horbachov, D. 2010. In the Epicentre of Abstraction: Kyiv during the Time of Kurbas. *Modernism in Kyiv.* Toronto: University of Toronto Press, 170-195.

Hryn, H. 2004. The Executed Renaissance Paradigm Revisited. *Harvard Ukrainian Studies*, 27(1/4): 67-96.

Ich.unesco.org. 2018. *UNESCO-Petrykivka decorative painting as a phenomenon of the Ukrainian ornamental folk art.* [online] Available at: https://ich.unesco.org/en/RL/petrykivka-decorative-painting-as-a-phenomenon-of-the-ukrainian-ornamental-folk-art-00893 [Accessed 15 Aug. 2018].

Kordysh, N. L. 1953. Stone Age Dwellings in the Ukraine. *Archaeology*, 6(3): 167-173.

Koroleva, E.A. 2011. Paleolithic art of the carpathian-dniester region. *Archaeology, Ethnology and Anthropology of Eurasia*, 39(1): 34-42.

Magocsi, P.R. 2010. *A history of Ukraine: The land and its peoples.* 2 ed. Toronto: University of Toronto Press.

Mudrak, M.M. 1981. Vadym Meller, Les Kurbas and the Ukrainian Theatrical Avant-Garde: Hello from Wave 477. *Russian History*, 8(1/2): 199-218.

Onyshkevych, L.M.L.Z. 1999. Exponents of Traditions and Innovations in Modern Ukrainian Drama. *The Slavic and East European Journal*, 43(1): 49-63.

Snikhovska, K. 2017. *The embroidery of Vyshyvanka. From traditional technique to contemporary technologies.* Kongsberg: University College of Southeast Norway.

Tunnels and Underground Cities: Engineering and Innovation meet Archaeology, Architecture and Art, Volume 1: Archeology, Architecture and Art in underground construction – Peila, Viggiani & Celestino (Eds)
 ISBN 978-0-367-46574-2

The Art Stations

F. Brenci
MC A - Mario Cucinella Architects, Bologna, Italy

ABSTRACT: Distributed along the lines 1 and 6 of the Naples Metro network, art stations are the result of extraordinary constructive and synergic dialogue between architects and artists, engineers and committee. Though the architectural design of the work is the constant dialogue and collaborative effort of all the players - namely architects, artists, engineers and Committee. The subway becomes an 'underground museum' and can be referred to as 'catacombs of beauty'. It is a 'traveling' art that supports the speed of the contemporary world and fulfils what its main features are. It becomes so familiar to the 'travellers': familiar and social, and goes beyond the concept of decoration, because, as complex and colourful it is, it strategically attracts the attention of the citizen, which is culled during his daily routine. More than 200 works are counted in Art Stations.

1 INTRODUCTION

'I would like people in general, and not only architects, to understand that architecture is not only what it looks like, but also what happens in it.' Bernard Tschumi in *The Manhattan Transcripts, 1976-1981*.

The origin of the Art Stations in the city of Naples dates back to 1995, when it was clear that, beyond purely engineering intervention, there could be spaces that would make the difference for the citizens and the world.

Distributed along the lines 1 and 6 of the Metro network, art stations are the result of willingness to completion by Metropolitane di Napoli; performing extraordinary constructive and synergic dialogue between architects and artists, engineers and committee. Though the architectural design of the work is strongly influenced by the underlying engineering, it is the constant dialogue and collaborative effort of all the players - namely architects, artists, engineers and Committee that continues to lead to the results we see.

Naples Metro Stations are a 'obligatory museum'. 'This is a confluence of beauty and transportation. We ask artists to create a piece to be inserted in the station' by Achille Bonito Oliva.

The stations are a 'obligatory museum', according to critic Achille Bonito Oliva, artistic coordinator of the subway company, because they invest in spaces that are a part of the citizen's daily life. At the price of a ticket commuters get to enjoy the works of international contemporary artists while on their journey.

The subway becomes an 'underground museum' and can be referred to as 'catacombs of beauty'. This is not a novelty for Naples, where the 'underground' element is a common feature and this distinguishes it from other cities. It still retains a precious archaeological heritage, which rises with every new project for a station. So, the 'underground' but contemporary museum connects with the tradition and the history of the city.

It is a 'traveling' art that supports the speed of the contemporary world and fulfils what its main features are. It becomes so familiar to the 'travellers': familiar and social, and goes beyond the concept of decoration, because, as complex and colourful it is, it strategically attracts the attention of the citizen, which is culled during his daily routine.

More than 200 works are counted in Art Stations. This highly technically demanding construction plan required often the embedding of enormous underwater boxes (100 m long and 35 m wide), with 3.5 m thick concrete walls.

Innovation is not limited to underground spaces, but it also involves the urban context. Installations are already present near the entrance to the subways, revamping the look of the neighbourhood that hosts it.

Planning a subway station inevitably means to establish a relationship between two opposite worlds: the city's surface, the everyday reality, and the underground, semi-unknown and unusual space for the life of the contemporary individual. Underground stations often become mere crossing points for hasty visitors, non-places as intermediate stages along a perpetual journey itinerary.

Following are some of the different approaches adopted by various architects and artists.

Entering the space created by Oscar Tousquets Blanca or Boris Podrecca means leaving the urban environment to get off in a silent and effective reality like that of the sea, a natural element for every Neapolitan. Both the Archistars have exploited the depths in which the subways are located to recreate the experience of entering the marine world, where it dominates the silence and away from the sunlight that filters through the urban portholes.

The University station is designed by architect and designer Karim Rashid, who offers a unique sensory experience. The artist investigated the world of young university where technology and research are expressed through colours, shapes and materials that allude to a different world, a modern world. It seems to be found in the world of the future, in which glittering colours dominate, such as pink and yellow and mirrored steel. There are words in the contemporary world such as 'virtual', 'network', 'database', 'interface' and 'software'.

The Garibaldi station, designed by the architect and urbanist Dominique Perrault, has the naked but luminous architecture that highlights the beams of light, which characterize the city's climate in a prevailing way. The interiors were made by Michelangelo Pistoletto, who intervenes with colourful representations in which the protagonists are the passers-by. Silhouettes on a mirrored stand represent common men and tourists in transit, but among them one can imagine the distracted passage, which is reflected in it. In this case the traveller enters the work and the latter takes on its value with interaction with the audience.

The ongoing development of art stations, Neapolitan institutional realities that begin to look at the contemporary as a resource and the contribution of technology create a blend that launches the city of Naples into new perspectives.

2 LINE 1

2.1 *Garibaldi*

Dominique Perrault, the French architect and urban planner of Garibaldi station, was also entrusted with the redesign of the square above it. The structure is conceived as a single, bright space traversed by spectacular intersections of the 'suspended' escalators.

The transparent glass roof allows natural light to reach almost the platform level, about 40 feet deep. The interior is strongly influenced by the choice of steel - satin or shiny and reflective - which is contrasted only by some bright orange elements.

The two installations by Michelangelo Pistoletto, one of the protagonists of the international art scene, are placed just before the last flights of stairs to the trains, one on the side of the arrival platform, the other on the platform towards Piscinola.

On steel mirror panels you can see full size silkscreened images of passengers, waiting or walking. Static images of art and ever-changing reflections from reality coexist incessantly in this work, which thus becomes, as explained by the author, 'a door connecting art and life.'

2.2 *Università*

A station offering a new sensory and aesthetic experience, characterized by soft volumes, vivacious florescent colours and innovative materials which can properly accommodate expressive and poetic needs; a transit area capable of touching the emotional sphere of the passenger, bringing beauty and pleasure into the daily commute - these were among the objectives at the heart of the Università Station-Line 1 Project of the unparalleled architect and designer Karim Rashid.

With university students and all metro users in mind, the Anglo-Egyptian architect envisioned spaces 'that embody the knowledge and language of the new digital age, that transmit the ideas of simultaneous communication, innovation and mobility, ideas which characterize their ongoing Third Technical Revolution'. Thus, from the moment they descend the stairs leading to the station, visitors find themselves surrounded by a multitude of words coined during the past fifty years, words such as 'virtual', 'network', 'operation', 'portable', 'database', 'interface', 'software' printed in pink and green on a ceramic background.

The ample and luminous atrium of the station lends a 'softening' friendly aesthetic which is spectacular at the same time thanks to the profusion of vivacious colours and digital images covering all wall and floor surfaces as well as the sensual qualities of the materials such as its perfectly smooth Corian walls or reflective steel of the vaults. But the colour contrast is also utilized to facilitate the circulation of passengers: the two dominant colours fuchsia-pink and lime green indicate the respective directions towards the platforms for Piscinola and for Garibaldi.

Beyond the turnstiles and Ticket Booth, on two large cylindrical pillars (Conversational profile), particular modelling of the volumes allows profiles of faces to be seen from all vantage points, a metaphor of dialogue and communication among human beings.Ikon, a sinuous sculpture in satin-finished steel, recalls human intelligence and in particular the synapses of our brain.

Moving towards the escalator leading to the Piscinola Direction platform, we are accompanied by light which springs from the translucent crystal ceiling panels, silkscreened in pink and light blue with images from Rashid's repertoire. The same decorative motif is repeated on the walls surrounding the descent. On the -1 mezzanine level, the colours of the flooring change from black with light blue, yellow and green flecks, as found in the atrium, to vivacious orange-pink tones. Pink is also found in the background of the circular light box in which bears a yellow cross.

Arriving on the -2 level, our attention is captured by the colours and shapes of the digital graphic motif which is also represented in the large floor tiles thus creating a fascinating pattern with three-dimensional effects in yellow, pink and light blue. In addition to the presence of two rectangular light boxes, on this level, one finds a spectacular surprise: at the height of the steps, giant images of Dante Alighieri and Beatrice - a homage of Rashid to the father of Italian literature.

2.3 *Municipio*

When work is complete, the Town Hall project undertaken by Portuguese architects Àlvaro Siza and Eduardo Souto de Moura will bring metro lines 1 and 6 together in one large transport hub. This new underground space will serve as a pedestrian link between the port and the historical area of the city, while urban planning at surface level will enhance the perspective axis from the Maritime Station at the Town Hall to the San Martino hill.

The Fountain of Neptune, a large marble group created by the work of Domenico Fontana, Michelangelo Naccherino and Pietro Bernini up till the end of the 1500s and subsequently by Cosimo Fanzago, has already been relocated to the square in front of the Town Hall.

The large Municipio transit area will incorporate and showcase the exceptional archaeological finds unearthed during the excavation work for the station in one of the largest archaeological investigations of the last decade. The antiquity unveiled by the excavations begins with the ancient port of Neapolis, which stood in a small inlet now buried under part of the present-day square. The excavations ass brought to light the sea bottom, sunken ships, a large wharf of the Augustan age, and traces of the occupation of the coast in the Hellenistic and Imperial ages. We continue with extraordinary remains of buildings of the Angevine period erected at the same time as Castelnuovo in the late thirteenth century, and the outer fortifications of the castle, built under king Alphonso V of Aragona and then the Spanish viceroys, with the towers of the Molo and the Incoronata. The latter is already visible today on the landing of the Line 1 station.

The station contains a single, large contemporary art installation: *Passages* by Michal Rovner (Tel Aviv, 1957), a 'video-fresco' - as defined by Achille Bonito Oliva, artistic coordinator of the Art Stations - in which images from five high-precision projectors blend with images drawn in pastel and painted with water colours by the artist directly onto the long, white wall of the atrium (37.70m × 4.00m).

2.4 *Toledo*

The project of the Catalan architect Oscar Tusquets Blanca also affected the area above, transformed into pedestrian zone and upgraded aesthetically.

The communication between internal and external space is entrusted to the skylight-structures-that, from the street, carry the sunlight in the rooms below.

On the first underground floor the remains of the walls of the Aragonese period are integrated into the architectural design, while the cast of a plowed field of the Neolithic, found during the excavations of the station is displayed at the Station Museum, in 'Stazione Neapolis' in the corridor connecting with the National Archaeological Museum.

In the coatings of this first level black predominates, an allusion to the asphalt of the contemporary city, which enhances the appearance of large mosaics by William Kentridge. The first is a long procession of dark figures, many of them inspired by the history of the city of Naples, led through music by the patron saint, San Gennaro. The background on which all the characters seem to pace slowly is the project for the *Central Railroad for the city of Naples, 1906 (Naples Procession)* which is also the title of the work. The second mosaic, located above the escalators, is titled *Remediation of the slums of Naples in relation to the railway station, 1884 (Naples Procession)*. This time the design used for the background of the work is the famous first project for a subway in Naples, created by the versatile Lamont-Young.

Going down to another level, upholstery colours change and we see a bright yellow reminiscent of the warm colours of the earth and the Neapolitan tuff, up to the level 0, or the sea level, indicated by the transition to spectacular mosaics of a blue which is becoming more intense as we go deeper.

This brings us to a monumental underground room, dominated by the charm of the oval mouth of the *Cráter de luz*, a large cone that crosses in depth all levels of the station, connecting the street level with the spectacular hall built 40 meters underground.

On the walls of the hall 'underground' we can admire the *Olas*, waves in relief designed by Oscar Tusquets Blanca, while proceeding within the tunnel overpass, we are surrounded by panels of the Sea by Robert Wilson, *By the sea ... you and me*, this is their title, light box with LED light made using the lens.

Men at work, a series of photos by Achille Cevoli on the walls near the fixed stairs, is dedicated to the theme of factory work, a tribute to those who made the excavation of the tunnels and the construction of the stations.

The second exit of Toledo station in largo Montecalvario is also enriched with works of art by internationally renowned artists.

Two long light-boxes by Oliviero Toscani run along the moving walkways connecting the two exits. The work, entitled *Razza Umana*, is part of a photographic study on the morphology of human beings. Many of the photos included in the Neapolitan installation, in some cases depicting the faces of public figures, have been shot in the squares of the city, others in other parts of Italy or the world, 'to see - as explained by the same author – how we are, what face do we have, to understand the differences. We take somatic fingerprints and capture the faces of humanity'.

On the walls above the long stairway that leads to the upper levels black panels are installed with mirror silver typefaces by the American artist Lawrence Weiner, one of the leading exponents of conceptual art, which has made of the graphic value of the word its privileged means of expression. *Molten copper poured on the rim of the bay of Naples* ('Rame fuso colato sulle rive del golfo di Napoli') is the epigrammatic phrase, in English and Italian, that Weiner offers us in this case and which also gives its title to the work.

Of an intense theatrical dramatic force is the installation of nine large portraits in black and white made by one of the most charismatic personalities of the contemporary scenario, Shirin Neshat, visual artist and filmmaker of Iranian origin. For this work Neshat has chosen for the first time in her career Western subjects, related to the environment of the Neapolitan theatre, particularly to the Teatro Nuovo which is located just a few steps from Montecalvario, and the Teatro Instabile. Among them we can recognize the actresses Cristina Donadio, Antonella Morea, Giovanna Giuliani and the artistic director of the Teatro Instabile, Michele Del Grosso. The title of the work – for which Shirin Neshat has partnered with the Neapolitan photographer Luciano Romano – is *Il teatro è vita. La vita è teatro - Don't ask where the love is gone* and explicits as much the inspiration to the correspondence relationship between the theatrical fiction and real life, as the desire to represent, through nine different expressions of the body, the feeling of loss and separation.

The Flying - Le tre finestre, by Ilya ed Emilia Kabakov, in ceramic of Faenza, is a large, airy panoramic vision that sees human beings soar in the sky with flocks of birds and airplanes.

For the station entrance level Francesco Clemente, among the leaders of the world's art scene since the 80s with the art movement of Transavantgard, has realized *Engiadina*, a spectacular work in mosaic and ceramic, more than sixteen meters long, depicting a mountain landscape crossed by a 'Clemente yellow' ceramic band, on whose background there is a parade procession of more than forty female figures, inspired by ancient Minoan era images of dancers found in the island of Crete. 'The title of my mosaic – said Clemente – refers to the Engadine valley in the canton of Graubünden in Switzerland, attended by the philosopher Nietzsche and the artist Segantini. I chose this valley because it is the last place where the Mediterranean light stops'. To realize Engiadina the Neapolitan artist, who has been living in New York for many years, has worked with the master potters of Vietri sul Mare and with Bruno Amman, specialist in stone mosaic, that, to achieve the full range of colours and light effects and the refined work, has selected more than one hundred different species of marble from around the world.

2.5 *Dante*

The station in Piazza Dante was inaugurated on March 27th 2002. The plans of the architect Gae Aulenti also concerned the redesign of the square with respect to its 18th century buildings. The Etnean stone flooring and cubic tiles follow the architectural designs of Vanvitelli, and the entrances to the station, in clear crystal and steel, were built in order to ensure visibility of the hemicycle from all sides.

The interior of the station is covered with large white glass panels with steel studs and house the works of some well-known figures of international contemporary art.

In the atrium of the station are two canvasses by Carlo Alfano; *Light-Grey*, from 1982, and *Fragments of an Anonymous Self-Portrait*, from 1985. Above the stairs which lead down to the lower level Joseph Kosuth, one of the fathers of conceptual art, places his work *These Visible Things*, made up of a passage from the *Convivio* of Dante Alighieri, 'written' with white neon tubes.

The wall of the lower level is occupied along its entire length by the *Untitled* of Jannis Kounellis: large steel panelling on which 'putrelle' similar to rails have been mounted, blocking numerous pairs of men and women shoes, a toy train, an overcoat, and a hat.

Continuing on down towards the platforms, above the escalators we find two versions of *Intermediterraneo* by Michelangelo Pistoletto, a reflective work which traces the outline of the Mediterranean Basin.

Lastly, *Universe without bombs, kingdom of Flowers, 7 red Angels* by Nicola De Maria, a long mosaic that stretches from the floor up to the ceiling, is a dance consisting of a multitude of small protruding geometric shapes and seven large colourful ovoids.

2.6 *Museo*

Built according to the designs of the architect Gae Aulenti and inaugurated in April 2001, the station is presented as a sequence of essential volumes of red plaster and Vesuvian stone that remind us of different street level, evoking through its colour and materials the nearby National Archaeological Museum. The interior, like in Dante station, is characterised by white glass facing and steel finishing.

The atrium of the station houses a fibreglass cast, created by the Naples Academy of Fine Arts, of the *Farnese Hercules,* while just inside the secondary entrance is a bronze cast of the monumental Horse Head called 'Carafa'.

Moving through the hallways towards the National Archaeological Museum, the black and white photographs of Mimmo Iodice anticipate the voyage to the ancient world with *Anamnesis* and with the series of *Athletes* and *Dancers*, works in which the Neapolitan master evokes the famous sculptures from the Villa of the Papyruses of Herculaneum and kept in the nearby museum.

In the upper entrance is located the bronze reproduction of Laocoön, that the Historic Chiurazzi foundry created based upon the ancient chalk cast kept in the Gipsoteca of the Naples Academy of Fine Arts. Behind the sculpture the large black and white photographs of Mimmo Jodice re-examine and enlarge details of the work, offering new suggestions for its interpretation.

The connecting corridor with the National Archaeological Museum houses 'Neapolis Station', the didactic exposition on archaeological sites discovered during the course of excavation work during the construction of Line 1, and in particular the stations of Municipio, Toledo; Università and Duomo.

The archaeological remains belonged to two primary settlements: Partenope, founded by Cuman colonists on Pizzofalcone Rock around the middle of the 7th Century before the common era, and Neapolis, erected between via Foria and Corso Umberto between the end of the sixth and beginning of the fifth centuries BCE.

The corridor that connects Museo Station with Cavour Station on Line 2 houses a selection of works by four artists, all from Campania and different generations, among the movers and shakers of modern photography.

Luciano D'Alessandro is present with nine works which highlight some the most significant moments of his career as a photographic journalist, characterized by a constant and empathetic attention to the human condition, from *Vendor of Small Paper Birds*, made in 1953, to *Cemetery of the Normandy Landings*, Saint Laurent, from 1994.

Proceeding down the corridor, we come across *India '70* by Fabio Donato, a series of shots taken during a youthful voyage in India. Also on display are three works which bear witness to close relationship between the Neapolitan photographer and Neapolitan art and theatre: *Be Quick*, with the gallery owner Lucio Amelio in front of the famous work by Andy Warhol, *Eduardo* and *Masaniello,* which captures a historic theatrical performance from 1976 starring Mariano Rigillo.

At the end of the artistic itinerary leading to line 2, there are photographs by Raffaela Mariniello, from which emerge the image of silent and motionless suburbs. *Hand-me-down Amusement Park Ride, On the Beach, Dresser*, and *Frame* isolate the particulars on which the focus lens has stopped, and these convey not so much a fragment of life, so much as an image gifted with its own autonomy and an accomplished formal organisation.

2.7 *Materdei*

Materdei Station was planned by Atelier Mendini, as was Salvator Rosa Station. With its opening in 2003, a new vitality and prestige was thus conferred on Piazza Scipione Ammirato, transformed into a pedestrian area, enriched with green spaces, new urban amenities and works of arts, such as *Carpe Diem*, the ironic sculpture in coloured bronze by Luigi Serafini and the ceramic reliefs which cover the external elevator, the work of Lucio Del Pezzo. Another interesting sight is the entrance which is covered by mosaic and presided over by a large green and yellow star.

The steel and coloured glass spire, quite similar to the ones at Salvator Rosa Station, accentuates the piazza and lightens up the entrance hall of the station, where green and blue tones predominate. The monumental mosaic by Sandro Chia with marine representations covers the base of the spire inside the station, while on a white wall are the striking solid geometric designs of Ettore Spalletti.

The ramp which leads down the lower levels passes underneath mosaics with ceramic reliefs by Luigi Ontani, a great marine expanse in which fantastic creatures and Neapolitan people splash around with a Pulcinella with the face of the artist.

On the track level, just after getting off the escalator, we find the refined drawings on wood panels by Domenico Bianchi, while the entire central corridor is covered by the extremely colourful *Wall Drawings* by Sol LeWitt, the father of minimal art, and creator of the fibreglass sculpture which is found at the end of the corridor.

Lastly, both platforms are enriched by the coloured silkscreens of Mathelda Balatresi, Anna Gili, Stefano Giovannoni, Robert Gliglorov, Denis Santachiara, Innocente and George Sowden.

2.8 *Salvator Rosa*

The station, planned by Atelier Mendini and opened to the public in 2001, was brought about through the close collaboration between architects and artists: the works of art, inside the extensive terraced gardens, dialogue with the architectural spaces and testimony of the past. The result, as noted by Alessandro Mendini himself, is a global aesthetic work that profoundly involves the citizen, and turns their daily life into a stage.

The area surrounding the station has benefited from the extensive revitalisation efforts which have returned the remains of a Roman bridge back its past splendour, as well as a graceful neoclassical chapel. It has also underlined the surrounding residential buildings, transforming them into works of art thanks to artist such as Mimmo Rotella, Ernesto Tatafiore, Mimmo Paladino, Renato Barisani and Gianni Pisani.

The different levels of the park are also interconnected by means of a long external escalator, which leads to games terrace, planned by Salvatore Paladino and Mimmo Paladino. On the ground, three playable games are made of volcanic stone inlayed with travertine marble: 'Tris', 'Bell', and 'Maze'. Calling us to play also are the light-hearted sculptures by Salvatore Paladino. In the same terrace, albeit in a more removed corner, is found the monumental 'hand' by Mimmo Paladino. The entire outdoors itinerary is punctuated by the works of many influential contemporary artists: Renato Barisani, Augusto Perez, Lucio Del Pezzo, Nino Longobardi, Riccardo Dalisi, Alex Mocika and Ugo Marano.

The station building itself is characterised by a fantastical eclecticism. The aboveground section, designed along simple yet forceful lines and faced with golden marble, evokes in the motif of the nearby Roman bridge in its succession of extensive arcades, while the steel and coloured glass spire transport us into a science-fiction future.

Along the walk from the atrium to the platform, it is possible to admire the works of Raffaella Nappo, Enzo Cucchi, LuCa, Santolo De Luca, Quintino Scolavino, Natalino Zullo, Perino&Vele and Anna Sargenti.

The station is also equipped with a second exit below via Salvator Rosa (opened in 2002), the presence of which is indicated by another spire by the Atelier Mendini, placed in the centre of a square. The base of the spire is covered in ceramic reliefs by Enzo Cucchi, refiguring a few icons of the Neapolitan imagination, while not far off, Lello Esposito's *Pulcinella*, with an outward curiosity 'looks out over the street, and observes the world and life'.

2.9 *Quattro Giornate*

The building of the station, which was designed by the architect Domenico Orlacchio and inaugurated in 2001, gave an entirely new look to Piazza Quattro Giornate. This included providing new gathering areas with green spaces. Besides these developments, there is a conspicuous continuity between the works of art located within the station, and those located in the area outside: the great metal sculpture of Renato Barisani, and the two bronzes of athletes by Lydia Cottone, placed between the garden's flowerbeds.

The station is- like the piazza surrounding it- named for the days of the uprising that freed Naples from Nazi occupation. The grand entrance hall houses paintings and reliefs in bronze by Nino Longobardi, inspired by the Neapolitan resistance. Descending towards the platforms, we find hunting scenes and "warriors" by Sergio Fermariello, the sculpture made of crushed sheets of aluminium of Baldo Diodato, and *Sabe que la lucha es cruel*, by Anna Sargenti.

The way up from the platform features three large reliquaries by Umberto Manzo, attached the wall by iron beams, a giant photographic image by Betty Bee confined to a light box, an oil on canvas by Maurizio Cannavacciuolo, called *Love against nature*, arriving finally at *Fighters* by Marisa Albanese, four white feminine sculptures which honour the resistance of the Four Days of Naples.

2.10 *Vanvitelli*

Planned by the architect Michele Capobianco and opened to the public in 1993, Vanvitelli station was restyled between 2004 and 2005 (thanks to Lorenzo and Michel Capobianco, as well as the artistic advice of Achille Bonito Oliva), renovating the station's large interior in order to house works by eight masters of modern art.

Luminous internal spaces are characterised by careful and sensible use of colours, ranging from blue to yellow, from lilac to various shades of grey.

The dazzling entrance hall features a work by Giulio Paolini, a giant boulder that appears to be on the verge of shattering its transparent enclosure. The two lateral corridors house on one side the long stripe by Vettor Pisani - an enigmatic and suggestive synthesis of images from differents epochs and styles - and on the other side photographs of the architecture of the city of Naples by Gabriele Basilico and Olivo Barbieri.

Unfolding upon the blue vault of the lower floor is the spectacular neon blue spiral of Mario Merz. The work- designed by the artist shortly before his death- runs along the vertical wall with a theme of prehistoric animals. Two large steel stars by Gilberto Zorio are attached to the side walls, filling the station's space with an interplay of fullness and emptiness.

The 'mouths of light' of Gregorio Botta, located at the intersection between the Garibaldi and Piscinola routes, are an invitation to slow down and have a look inside of the eight cylinders. On the platform floor, the two large mosaics by Isabella Ducrot draw the attention of the travellers to the chromatic and sensory qualities of the materials used in its execution.

2.11 *Rione Alto*

The station's second exit was inaugurated in December 2002, joining the art stations through the presence of the works of renowned international artists and those of young emerging Neapolitan artists. The outside layout, with glass and metal cupolas corresponding to each entrance, is further enriched by a mosaic by Achille Cevoli.

In the entrance hall is *Wall drawings* by David Tremlett, with their geometric forms running the length of the walls. Then, between the conveyor belts and the escalators, travellers come across seven panels by the Neapolitan Giuseppe Zevola. This is then followed by a sequence of obsessively reiterated faces by Katharina Sieverding and *Rem* and Jsr, the two light-boxes of the couple Bianco-Valente (Giovanna Bianco and Pino Valente), which loom over the traveller from the tops of the vaults of the gallery leading to the tracks.

After finishing the descent, the footballers in action by Marco Anelli accompany the traveller to the platforms, in addition to the permanent exposition of emerging artists selected by Paola Guadagnino: Pennacchio Argentato, Donatella Di Cicco, Danilo Donzelli, Pina Gigi, Ivan Malerba e Marco Zezza.

3 LINE 6

3.1 *Mergellina*

Mergellina station, designed by Studio Protec, opened to the public in February 2007. The station is particularly noteworthy for its slanting lift, which runs parallel to the escalator.

The entrance hall, designed by Vittorio Magnago Lampugnani, is a large rectangular space, with sidewalls covered by two enormous mosaics by Gerhard Merz, which seem to announce the passage from the station's interior to the light of open space through its light and 'atmospheric' colours.

The gates are the work of Alan Fletcher, an important figure in international graphic arts, who recently passed away. With his usual elegant simplicity, the English artist created a pattern through the intersection and repetition of the words 'metropolitana mergellina', carved into the metallic surface of the gate.

3.2 *Lala*

The building housing the Lala station is covered in travertine slabs, and like the external layouts, was designed by Studio Protec, (Uberto Siola, Luigi Milano, Luigi Pisciotti, Dante Rabitti, Federica Visconti). Located alongside Largo Lala, the building perfectly complements the circular perimeter of the piazza.

The interior of the station houses the works of five modern photographers. The Brazilian Salvino Campos is present with *Untitled 12/La Habana 2002*, centred on an immobile antique car, and *Capoeira, Salvador, Bahia 2004*, which captures the dynamism of a young male body engaged in an old South American dance.

The human form is praised in *Femme Terre, 1998-1999*, by the Senegalese Ousmane Ndiaye Dago: his female nudes, covered with a clay-like layer that makes them resemble living sculptures, recall the harmonious composedness of classical figures through the movements of their shoulders and their raised arms.

At the end of the narrow corridor leading to the platform for Mostra, we see a work which exerts a somewhat disturbing influence on the viewer: a screaming woman wearing a veil portrayed by the Neapolitan photographer Monica Biancardi.

In both the images produced by Luca Campigotto and Vicenzo Castella, the frames are dominated by industrial landscapes frozen in time, focusing our gaze on the current urban situation in Naples.

The only artist represented at Lala through an artistic installation in Nanni Balestrini, with his Allucco, an explosion of splinters, mirrors, and fragments of words.

3.3 *Augusto*

Augusto Station, designed by Studio Protec (Uberto Siola, Luigi Milano, Luigi Pisciotti, Dante Rabitti, Federica Visconti), and opened to the public in February 2007, is a building faced with sections of circles, looking out over Largo Veniero with a row of pillars.

The first works that one encounters when descending towards the trains for Mostra are the ceramic reliefs of Luisa Rabbia. In one of the corridors leading to the platforms, Franco

Scognamiglioinstalled a series of light boxes dedicated to the life of Galileo. Proceeding onwards into the interior of this "theatre of the memory" of sorts, the travellers are progressively engaged by the strong symbolic nature of the environment surrounding them.

The dramatic atmosphere of the city's outskirts come to life in the ambient installation by Botto&Bruno, which fully occupies the other access corridor leading to the platform.

If we come down through the entrance towards the Mergellina train, we encounter *The Milky Way*, a lively mosaic with a relief in ceramic by Cristina Crespo. Meanwhile, the platform floor features mosaics by Carmine Rezzuti and Matteo Fraterno. The first depicts, against the backdrop of a fiery sky, a roaring black panther guarding Mount Vesuvius; in the second a red vortex imposes a lively spinning movement to the entire work.

3.4 *Mostra*

Travellers are greeted by the black and white photographs of Gabriele Basilico, whose shots render homage to the monumental splendour of the architecture of the Mostra d'Oltremare.

In the large entry hall, three large mosaics made by Costantino Buccolieri are flanked the work of one of the greatest masters of 19th century Italy, Mario Sironi (Sassari 1885 – Milano 1961). Furthermore, in the corridor which connects the station with the Cumana line, travellers can admire the photography of Pino Musi. Alongside the three stairways that connect the entry hall to the platform floor, we come across *Monumento a G.P.*, in which the Neapolitan Gianni Pisani develops the theme of 'Suicide of the Artist', the large feminine face drawn by Marisa Merz, and the abstract composition of Carla Accardi.

4 CONCLUSIONS

The city is an 'event', reading the city as a big event composed of sequence of daily events where each space is designed or spontaneously become an occasion for an event to take a place.

Events takes a place in between the edges of 'architecture', reading the city as sequence of rooms and corridors where each space is articulated and expressed by architecture as a language and scale, therefore the uniqueness of the city is perceived.

Reading the city as a structure of values where each space or a landmark is an added 'value' to the city and represents the whole city experience for inhabitants and visitors. Reading the city as an 'infrastructural' organism, where the network of underground mobility became the skeleton of cities, preforms in the daily life experience and reshapes the users mental map. The underground network apart from the mobilization as a function it is sequence of spaces, events and gather points. And from here the question is raised; what are these new spaces to the 'city'?

Tunnels and Underground Cities: Engineering and Innovation meet Archaeology, Architecture and Art, Volume 1: Archeology, Architecture and Art in underground construction – Peila, Viggiani & Celestino (Eds)

The outlets of the Albano and Fucine Lakes and their influence from Roman times up to the birth of modern tunnel engineering

C. Callari
University of Molise, Campobasso, Italy

ABSTRACT: The historic background and the engineering aspects of the construction of the outlet tunnels of the Albano (398-397 BC) and Fucine (10 BC-54 AD) lakes are employed to show how the major underground works of Roman civilization served also as carriers of technical knowledge and organizational skills over the centuries, up to the birth of modern tunnel engineering in the nineteenth century.

1 INTRODUCTION

Water tunnels are the most extensive underground works constructed in Roman times. Besides the several kilometers of intakes and underground sections of the aqueducts, there are also some important outlets, such as those of the Albano (398-397 BC) and Fucine (10 BC-54 AD) lakes (Figure 1). These two outlets have been considered for many centuries as unbeatable models for tunnel engineering, as demonstrated by the several technical descriptions and artistic representations by important theorists of architecture and engineering, such as Giovanni Battista Piranesi, who made several engravings of both the outlets in the eighteenth century (Piranesi 1761, 1762, 1787).

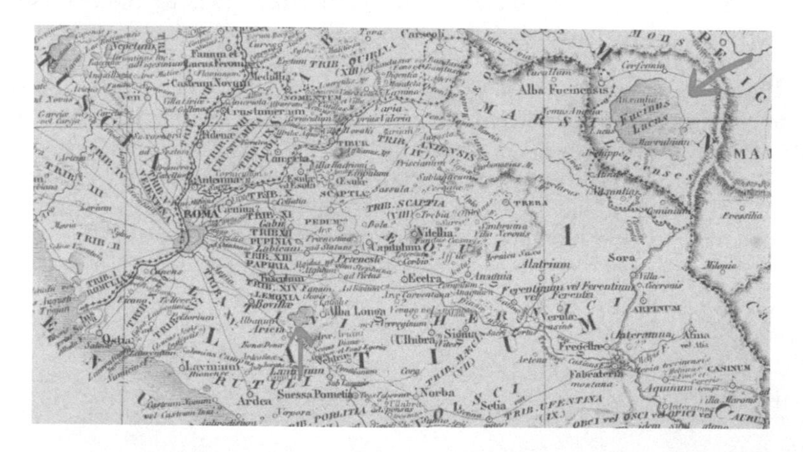

Figure 1. Location of the Albano and Fucine lakes in a map of the "Prima Italiæ antiquæ regio" (1851).

2 THE OUTLET OF LAKE ALBANO

The volcanic Lake Albano lacks a natural outlet (Figure 2). According to Titus Livius (27-9 BC), the outlet tunnel was excavated between 398 and 397 BC to regulate the lake level (Figure 3), which was experiencing an extraordinary raising, with great concern for the people living along the banks and downstream of the crater.

The main data of the outlet are summarized in Table 1. Despite its length, the tunnel alignment is quite smooth, as observed in various speleological explorations (Castellani 1999). It is thus surprising that such result was obtained with just two shafts, since a typical function of such structures was to facilitate the fulfillment of the planned tunnel alignment, starting from its projection on the ground surface. In Roman times, further main functions of the shafts were (and still are in modern tunnel engineering): opening of additional excavation faces; ventilation; mucking; delivery of construction material; exploration.

The outlet tunnel, entirely excavated in pyroclastic tuff, is unsupported for its whole length. As concerns the doubts often raised about the construction time, it can be remarked that in the cited speleological studies it was also observed the presence, on the tunnel walls, of thin vertical

Table 1. Main engineering data of the Lake Albano outlet tunnel.

Construction period	Length (km)	Face area (mean, m^2)	Ground type	Cover height (max, m)	Slope (mean)	Advance rate (mean, m/day)	Shafts (n.)	Inclined adits(n.)
398-397 BC	1.45	1.0 x 2.5	Pyroclastic tuff	120	0.14%	4.0	2 (known)	1 (assumed)

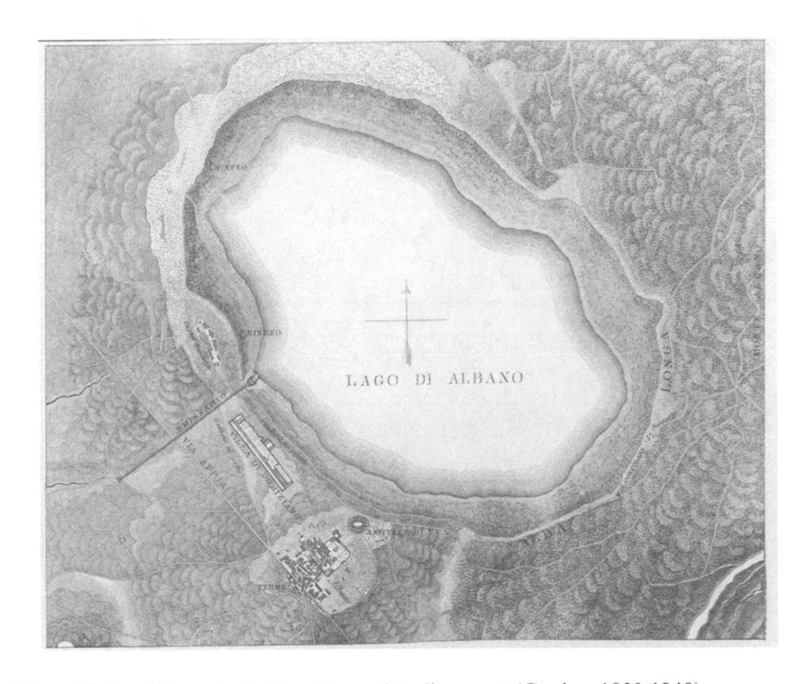

Figure 2. Plan of Lake Albano including the outlet alignment (Canina 1830-1840).

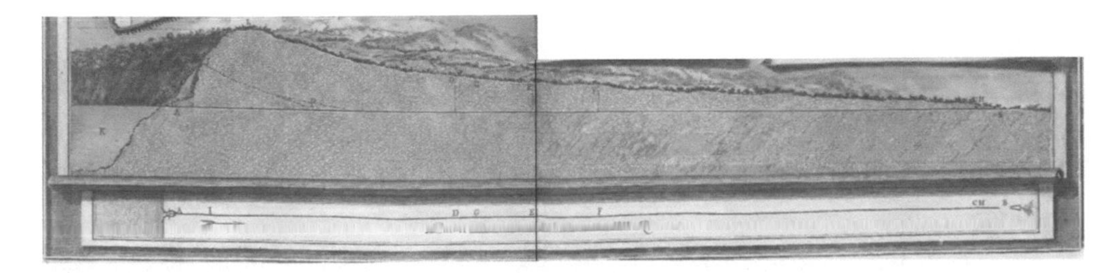

Figure 3. Longitudinal profile and plan of the Lake Albano outlet alignment (Piranesi 1762).

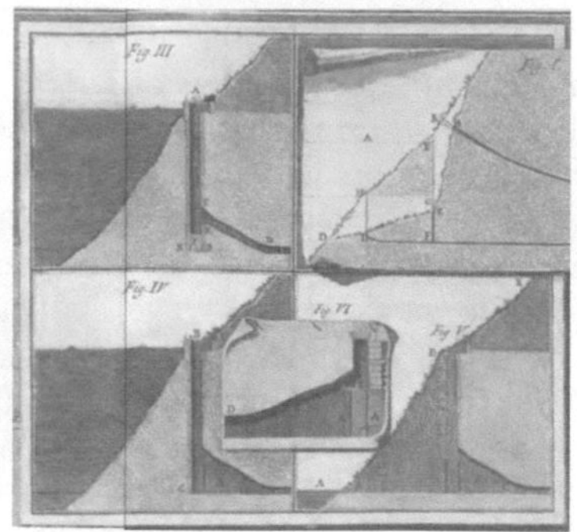

Figure 4. Intake of Lake Albano outlet: longitudinal profile (left) and conjecture about the relevant excavation stages (Piranesi 1762).

edges of protruding rock, placed at an average distance of 1.35 m. If we assume that such rock "frames" were used to mark the excavation face advancement in each single work shift, and that in the 24 hours there were 3 shifts of 8 hours each, the average daily advancement would have been of about 4 m. Furthermore, if the advancement of a single excavation face is reasonably assumed for most part of the tunnel length, the outlet construction would have lasted about one year. Such estimate is thus consistent with data reported by Titus Livius. It can be remarked that also the Piranesi engravings (1761, 1762), although affected by unrespected proportions and inaccuracies are reasonably consistent with our current knowledge (Figure 4).

The outlet efficiently operated at least up to the 1970s. Then, the lake level progressively lowered to below the elevation (293 m a.s.l.) of the outlet intake, namely, the lake level reached 283 m a.s.l. in 2007 (Anzidei & Esposito, 2010).

3 THE OUTLET OF LAKE FUCINE

Lake Fucine, dried up in the second half of the nineteenth century, was the third largest lake in Italy (maximum area: 170 km^2, length: 19 km, width: 10 km, maximum depth: 22 m). Due to missing outlet rivers, the lake was regulated exclusively by karstic ponors and evaporation. Therefore, the water level was very sensitive to climatic variations. Such highly irregular regime caused frequent flooding, with severe damage to cultivated fields and buildings (Pantaloni et al. 2016).

The first project of an artificial outlet was promoted by Julius Caesar, to control the lake level and obtain new cultivable areas. However, after his death, the project was abandoned by Augustus, perhaps because of the necessary financial commitment, as well as by his successors, Tiberius and Caligula, who did not carry out such a project. It was the emperor Claudius (10 BC-54 AD) who, on the contrary, again considered the construction of an outlet tunnel as a priority enterprise. The main motivation was, perhaps, of political nature: to supply Rome with further food, produced in a new cultivable vast extension of land, could have prevented the repetition of already occurred revolts of the plebs, caused by famines, which had posed a serious threat to the imperial authority (Brisse & De Rotrou 1876).

We do not have information about the designers, but we know that the main contractor was the freedman Tiberius Claudius Narcissus, who played a crucial role in the ambitious program of infrastructural expansion launched by Claudio. According to Roman historians, Narcissus would take advantage of this and other contracts to improperly take possession of an enormous wealth (Tacitus 1st cent. AD).

3.1 *Roman outlet: geomechanical profile, shafts and adits, excavation and support techniques*

The work, hereafter referred to as the "Roman outlet", was carried out by means of a tunnel excavated under Mount Salviano and Palentini Fields, up to reaching the Liri River

Figure 5. Plan of the (practically coincident) alignments (ABCD) of the Torlonia and of the Roman outlet tunnels of Lake Fucine. The lake is on the right (D) and the Liri River is on the left (A) (Brisse & De Rotrou 1876).

(Figure 5). Claudio's tunnel was excavated through sound limestone, clays with sand intercalations, conglomerates, detritus (Figure 6), i.e. in much more difficult conditions than those experienced with regard to the Lake Albano outlet.

Among the main data of the outlet, listed in Table 2, we remark those related to the great commitment in terms of auxiliary underground works (shafts and adits, Figures 6–8), which

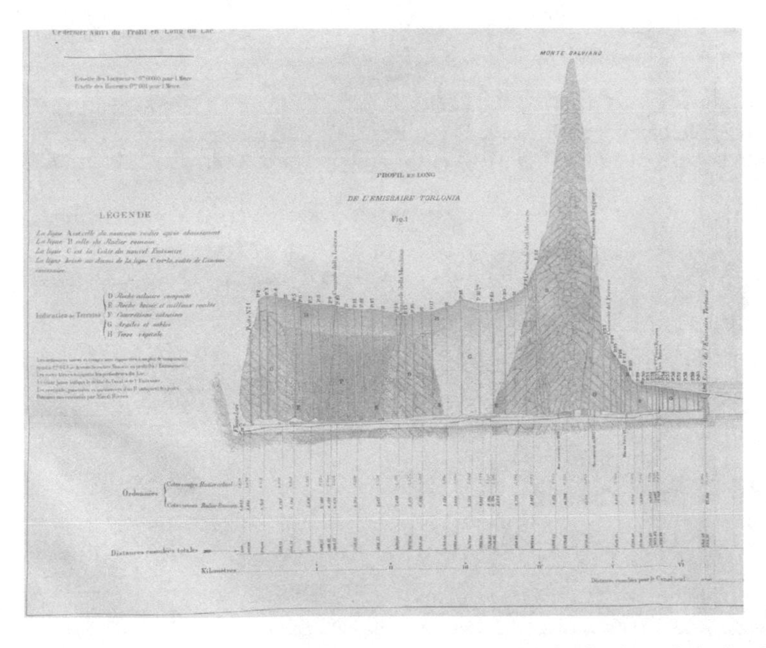

Figure 6. Longitudinal profile of the outlets of Lake Fucine with indication of shafts, inclined adits (called "cunicoli") and encountered soil and rock masses: D) limestone; G) clays with sand intercalations; E) detritus; F) conglomerates; H) topsoil. Invert and crown of the Torlonia outlet are marked in red. The increment of the tunnel volume obtained by means of lowering excavation from the Roman outlet is highlighted in yellow. Upstream, the Roman outlet stopped at the progressive "Roman tanks". The ratio between vertical and horizontal scales is 20 (Brisse & De Rotrou 1876).

Table 2. Main engineering data of the Lake Fucine outlet tunnels.

	Constr. period	Length (km)	Face area mean (m^2)	Ground type	Cover height (max, m)	Slope (average)	Advan. rate (average) (m/day)	Shafts n. depth (m)	Inclined adits (n.)
Roman outlet	41-52 AD	5.7	5.1	limestones clays & sands conglomerates detritus	from 100 to 300	0.15%	1.4	≥33 18 ÷ 122 m	≥12
Torlonia outlet	1854-1869	6.3	19.6				1.2	28 restored	2 new

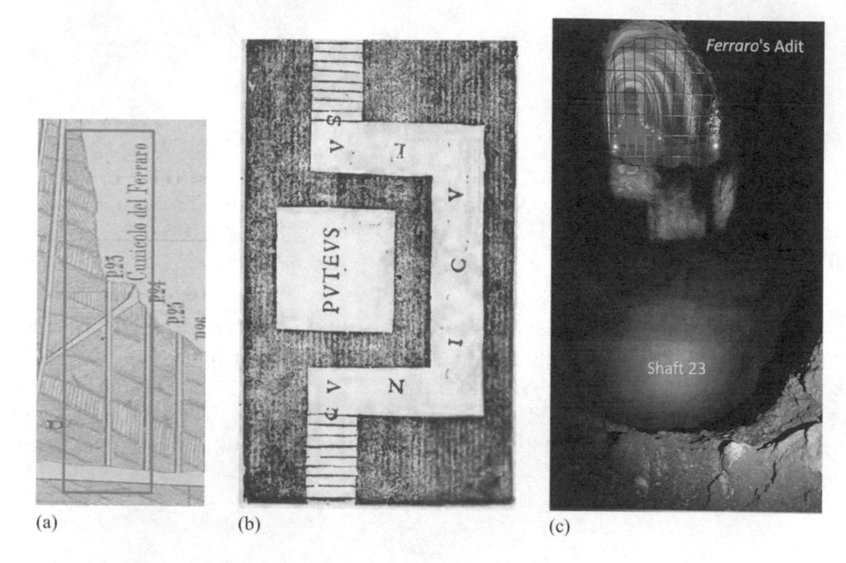

Figure 7. *Ferraro*'s Inclined Adit (*Cunicolo del Ferraro*) and Shaft 23 serving the Lake Fucine outlet: a) longitudinal profile in limestones (Brisse & De Rotrou 1876); b) plan of their intersection (Fabretti 1683); c) shaft and adit seen from the upper portion of the adit on April 2018.

Figure 8. Lake Fucine outlet: a) detail of a Roman engraved stone found near the *incile* (intake) which depicts, behind the turbulent lake waters crossed by ships, two vertical capstans turned by men and serving two shafts; b) Tools for excavation, wooden carpentry and mucking found in shafts 14 and 16 (Marsica Museum, April 2018). The engraved stone and the tools were found by Afán de Rivera (1836).

played a crucial role in the construction of the main tunnel. Indeed, the total length of such auxiliary structures exceeds twice that of the water tunnel. In this regard, we remark that in Table 2, the average daily advancement is calculated considering only the water tunnel, i.e. it does not account for shafts and adits. On the contrary, the average daily production for the whole system of underground works was almost 4.3 m/d. Suetonius (1st cent. AD) reports that 30,000 workers would have worked at the outlet construction. Such a number was considered as realistic by those engineers who directed the works of rehabilitation (Afán de Rivera 1823, 1836) and upgrading (Brisse & De Rotrou 1876) of the outlet. However, a simple calculation based on data in Table 2 and on the optimistic assumption of simultaneous excavation at all possible fronts with a reasonable duration of the workshifts, shows that the number of workers mentioned by Suetonius is at least 10 times larger than a realistic value.

According to recent speleological explorations (Burri 2005) the final cross section of the tunnel was obtained by widening exploratory adits (about 80 cm wide) which link adjacent shafts. In the sections excavated in clays with sand intercalations and in conglomerates, the tunnel was supported by a final concrete lining, with brick masonry abutments (Figures 9–10). Based on direct visualization, Brisse & De Rotrou (1876) provide an accurate description of the techniques adopted by the Romans for temporary supports, using wooden props, and the aforementioned final concrete lining. In this way, we know that the Romans incorporated props inside the concrete casting, cutting their protruding portions from the vault intrados. Over the centuries, the degradation of wooden props caused the formation of voids in the concrete, with subsequent weakening of the structure and widespread water inflows. In the Roman tunnel, the presence of voids was detected also between the vault extrados and the soil, together with the subsequent effects of uneven distributions of the interaction pressure.

3.2 *Instability phenomena faced by Romans during outlet tunnel construction*

The geotechnical difficulties faced by Romans during the outlet construction can be inferred from the strong variability of the tunnel cross-section size along the alignment. In Figures 6, 10, it is observed that in the sections characterized by poor geotechnical properties (e.g. in clays with sand intercalations and in detritus), the area of the excavation face was significantly reduced with respect to its maximum value, to increase the stability of the underground opening.

Furthermore, from the alignment deviation (JQCDEGF) shown in Figure 11, detected during the works of rehabilitation (Afán de Rivera 1823, 1836) and upgrading (Brisse & De Rotrou 1876) carried out in the nineteenth century, it can be inferred that the most difficult situation for the Romans occurred between shafts 19 and 20 in Figure 6, i.e. in the Quaternary age fluvial-lacustrine deposits of the Palentini Fields, near the limestones of Mount Salviano. From the cited descriptions, it can be understood that the tunnel collapse was caused by the poor strength of the clays and, more importantly, by the seepage forces triggered in sand layers by excavation. The

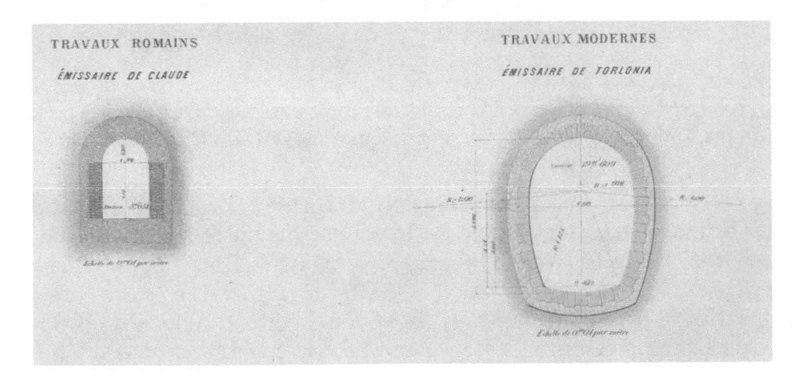

Figure 9. Typical tunnel cross-sections of the Roman (left) and of the Torlonia (right) outlets. Except for the masonry brick abutments, the final lining of the Roman outlet is of concrete. The Torlonia tunnel is supported by a hewn stone masonry (Brisse & De Rotrou 1876).

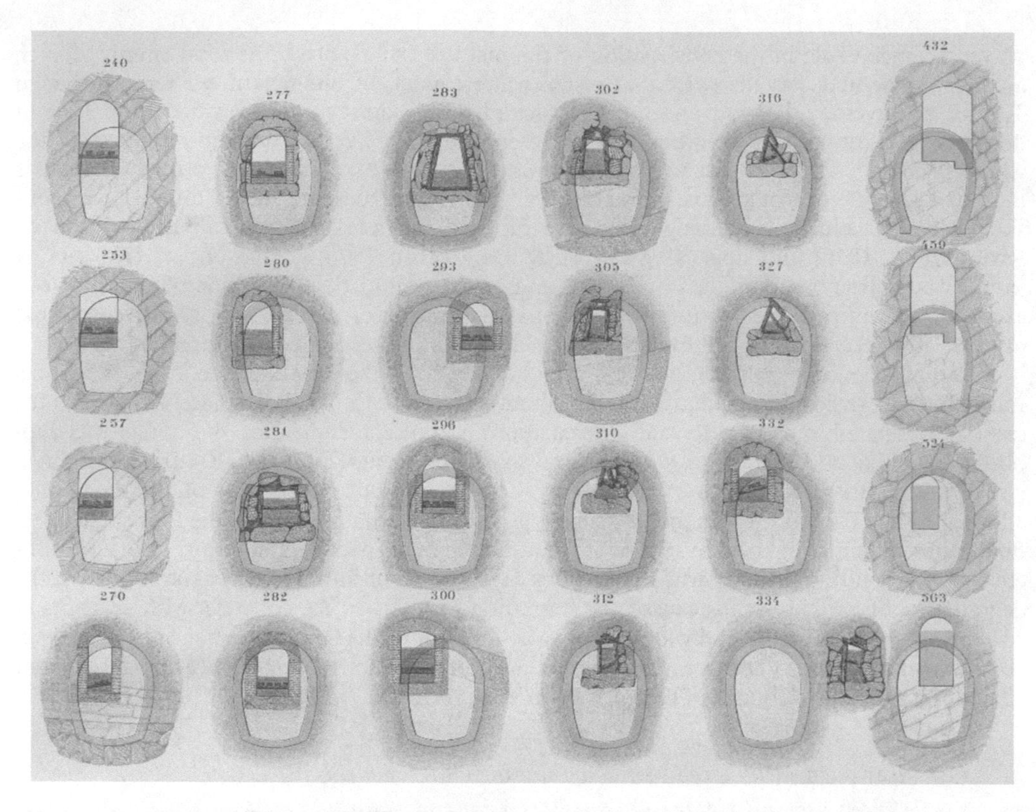

Figure 10. Cross sections of the Torlonia and of the Roman outlets at different chainages (given in decameters and increasing from downstream to upstream). The Roman tunnel is the smaller and is typically located in the upper portion of the Torlonia outlet (Brisse & De Rotrou 1876).

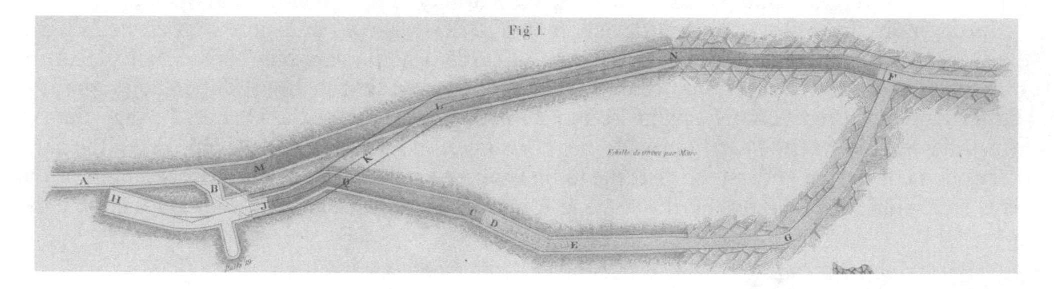

Figure 11. Plan view of the section (MLNF) of the Roman outlet, located between shafts 19 and 20 (see Figure 6), which collapsed during construction, and subsequent deviation JQCDEGF. The JQC section collapsed in 1842. In the Torlonia outlet, the final tunnel alignment is AHJLNF (Brisse & De Rotrou 1876).

collapse affected a significant tunnel length (MLNF in Figure 11) with subsequent outlet obstruction. The water of the lake, no longer able to flow, filled the tunnel portion located upstream of the collapsed section. It can be assumed that the Romans resolved such a difficult situation by excavating from shaft 19 along the deviation JQCDEGF (Figure 11), i.e. until they rejoined the original alignment in F, far enough inside of the sound limestone. Since the upstream portion of the outlet was filled with pressurized water, it can be also conjectured that the breakthrough of the rock diaphragm in F was anticipated by a progressive draining through holes made in the same diaphragm. The collapsed MLNF section was then closed with a wall, while the JQCDEGF deviation became part of the Roman outlet alignment until 1842, when it was obstructed by the collapse of the JQC section.

A debated issue is related to the Roman outlet operation, which did not lead to the complete dewatering of the lake but reduced its surface by about two thirds. According to a first hypothesis, such result would have been caused by mistakes in leveling, hydraulic design and construction phases. According to another conjecture, the drying up of the lake was not Claudio's objective, also as a sign of respect for the indigenous Marsi people, likely scared by a possible revenge by God Fucine. In favor of the latter hypothesis, we mention the arguments by Brisse & De Rotrou (1876), aimed at demonstrating, using their geometric survey of the ancient outlet, that the engineering and technological skills of the Romans were sufficient to achieve a complete dewatering of the lake.

3.3 *Rehabilitation and upgrading works in the nineteenth century*

The outlet effectively regulated the lake level, preventing floods, at least until the VI century AD, when the lack of maintenance and perhaps a seismic event (Giraudi 1988) stopped its operation. In the centuries that followed, there were several but unsuccessful rehabilitation attempts (e.g. that of Frederick II in 1240).

In the XVII and XVIII centuries, many prominent experts in architecture and engineering studied the Roman outlet. Among others, we mention the antiquarian Raffaele Fabretti (1683, Figures 7b,12) and Piranesi (1787, Figure 13). Also the *Premier Architecte du Roi,* Robert De Cotte (1685), studied the ancient tunnel in order to obtain indications for the design of new aqueducts, canals and sewers to be constructed in France.

In 1815 the lake level reached the maximum recorded elevation (corresponding to a maximum depth of 23 m (Brisse & De Rotrou 1876)) causing a strong concern among the populations of Fucine. For this reason, Carlo Afán de Rivera (1779-1852), who was the director of the "Corpo di Ponti e Strade, Acque, Foreste e Caccia" of the Kingdom of the Two Sicilies, was charged with the rehabilitation of the Roman outlet. Although with considerable difficulty, Afán de Rivera (1823, 1836) succeeded in desilting and strengthening most of the shafts and inclined adits. Conversely, the Neapolitan engineer failed in the full rehabilitation of the main tunnel, because of the hesitations displayed by the Borbonic administration and of major geotechnical difficulties. Among the latter, we mention the catastrophic collapse, dated 1823, of the tunnel crown in the conglomerates near the "Lamp Adit" (*Cunicolo della lucerna* in Figure 6). The size of the collapse

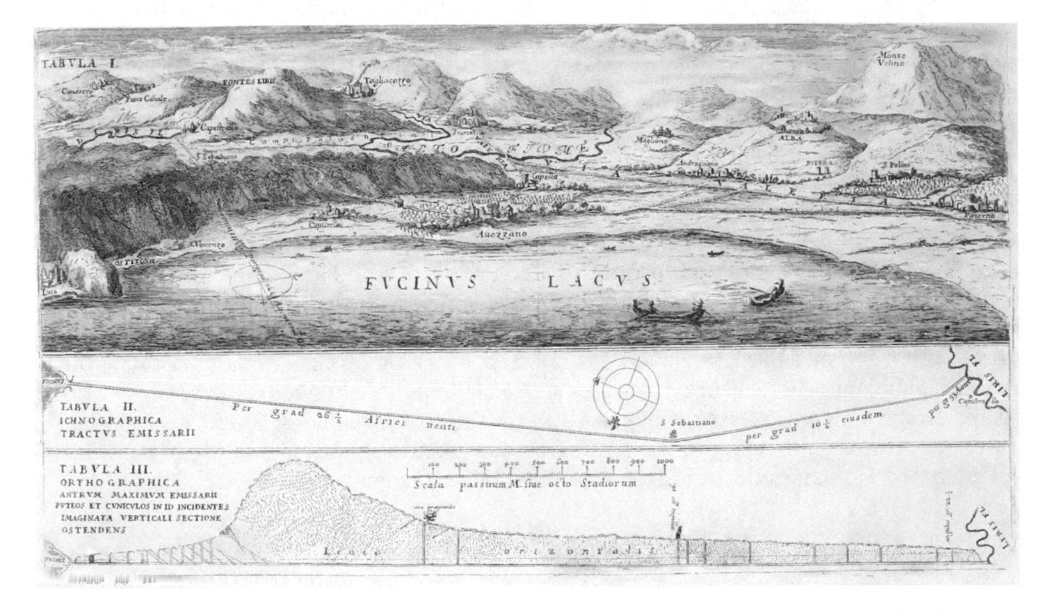

Figure 12. Roman outlet of Lake Fucine (Fabretti 1683): view of the lake with tunnel alignment ("Tabula I"); plan ("Tabula II"); longitudinal profile including shafts and adits ("Tabula III"). These and other Fabretti's drawings are also attached to the study by De Cotte (1685).

Figure 13. Roman outlet of Lake Fucine. "Fig. XII": internal view of the outlet tunnel (A) with one of its inclined adits (B). "Fig. XIII": cross section of outlet and adit shown in Fig. XII (Piranesi 1787). In the latter engraving, the reported tunnel size is realistic (in limestone). On the contrary, the height of human figures is very low (about 1.35 m), likely to give the impression of a larger tunnel (see also Fig. XII).

mechanism was such that its effects propagated up to the surface, with the formation of a sinkhole, 20 m large and 10 m deep (full of water for 8 m). In spite of this, Afán de Rivera (1836) was able to advance through the collapsed section by supporting the face with propped wooden boards. Each single board was in sequence removed to allow for a localized excavation advancement of 40 cm; the board was then propped again versus the face. We remark how such procedure was similar to the poling board technique employed some years later by M.I. Brunel in the excavation of the Thames Tunnel (Muir Wood 1994). In the same tunnel section, Afán de Rivera (1836) employed also a pioneering forepoling technique.

In 1852, a Borbonic decree assigned to a Neapolitan company the rehabilitation and upgrading of the Roman outlet, with the commitment of the complete dewatering of the Lake Fucine. In 1854, the stock actions of this company were acquired by Duke Alessandro Torlonia (1800-1886), who, about the related financial risk, commented with worried irony: "Either I dry the Fucine, or the Fucine dries me". The works of the Torlonia outlet started in 1855 under the direction of the famous Swiss engineer, naturalized French, Jean François Mayor de Montricher, who died of typhus in Naples in 1858. His successor, the Swiss engineer Henri Samuel Bermont, also had to abandon the appointment in 1869 for serious health problems (he died in 1870). The outlet tunnel was thus completed in 1869 by the French engineer Alexandre Brisse, assisted by the administrator Léon De Rotrou. The complete drainage of the lake was reached in 1875. The works of the outlet were completed in 1876. In the same year, Torlonia was appointed Prince of Fucine by the Italian King Vittorio Emanuele II.

As summarized in Table 2, the main differences between the Roman and the Torlonia tunnels are: the significant enlargement of the cross section (Figure 9), obtained by lowering excavation starting from the Roman outlet (Figures 6,10); the improved quality of the final supports, consisting of hewn stone masonry in the Torlonia outlet (Figure 9); the elimination of the deviation existing in the Roman outlet, using the rectified AHJLNF alignment in Figure 11.

4 CONCLUSIONS

The outlets of lakes Albano and Fucine can be considered as paradigms of the crucial role played in Italy by large underground works since classical antiquity, not only as public infrastructures, but also as carriers of technical knowledge and organizational skills up to the birth of modern tunnel engineering in the nineteenth century. The centuries-old story of Fucine outlet and the inclusion by Piranesi (1761) of the Lake Albano outlet among the representative works demonstrating the "Magnificence and Architecture of the Romans" are examples of the cultural influence of these tunnels, which lasted far beyond their very long operation time.

From the documents drawn up by designers and project managers of the nineteenth-century works in Fucine, it can be inferred a technical knowledge increasing at impressive rate within that century. In such documents, it is also observed a certain inclination, still common today, to charge others with liability for delays, accidents and unpredicted expenses. It is for example the case of Brisse & De Rotrou (1876), who often attribute the liability for delays and accidents occurred during their contract to previous mistakes by the Romans and Afán de Rivera.

ACKNOWLEDGEMENTS

The author is grateful to Paolo Lupino, Flavia Baldassarre and Carlo Grillo of the *Parco Regionale dei Castelli Romani* and to Gino Di Berardino, President of the *Consorzio di Bonifica Ovest del Liri-Garigliano*.

REFERENCES

Afán de Rivera, C. 1823. Considerazioni sul progetto di prosciugare il lago Fucino e di congiugnere il mar Tirreno all'Adriatico per mezzo di un canale di navigazione. Napoli: Reale Tipografia della Guerra.
Afán de Rivera, C. 1836. Progetto della restaurazione dello scolo del Fucino. Napoli: Stamperia e Cartiera del Fibreno.
Anzidei M. & Esposito A. 2010. The lake Albano: bathymetry and level changes, in Funiciello, R. & Giordano, G. (ed.), The Colli Albani Volcano. Special Publication of IAVCEI, 3. London: The Geological Society, 229-244.
Brisse A. & De Rotrou L. 1876. Desséchement du lac Fucino exécuté par S.E. le Prince Alexandre Torlonia. Précis historique et technique/The draining of lake Fucino accomplished by His Excellency Prince Alexander Torlonia. An abridged account historical and technical. Rome: Imprimerie de la Propaganda/Propaganda Press.
Burri, E. 2005. Il Fucino e il suo collettore sotterraneo. Opera Ipogea 1(2): 56-74.
Canina, L. 1830-1840. L'architettura romana, descritta e dimostrata coi monumenti. Rome: Tipi dello stesso Canina.
Castellani V. 1999. Civiltà dell'acqua. Rome: Editorial Service System.
De Cotte R. 1685. Dessins de la décharge du lac Fucino, dans la rivière de Gariglian. . .avec explications.
De Rotrou L. 1871. Prosciugamento del lago Fucino eseguito dal principe Alessandro Torlonia. Firenze: Tipi dei successori Le Monnier.
Fabretti R. 1683. Emissarii lacus Fucini descriptio. Rome: Nicola Angelo Tinassi.
Giraudi G. 1988. Evoluzione geologica della piana del Fucino (Abruzzo) negli ultimi 30.000 anni. Il Quaternario 1(2): 131-159.
Muir Wood A.M. 1994. The Thames Tunnel 1825-43: where shield tunnelling began. Proceedings of the Institution of Civil Engineers - Civil Engineering 102(3): 130-139.
Pantaloni M., Console F. & Perini P. 2016. Evoluzione della Piana del Fucino tra bonifica e terremoto (1870-1915), Terremoti e altri eventi calamitosi nei processi di territorializzazione, Roma: Labgeo Caraci.
Piranesi, G.B. 1761. Della Magnificenza ed Architettura de' Romani. Roma.
Piranesi, G.B. 1762. Descrizione e disegno dell'emissario del Lago Albano. Roma.
Piranesi, G.B. 1787. Dimostrazioni dell'emissario del Lago Fucino: alla Maestà di Fernando IV Re delle Due Sicilie e di Gerusalemme.
Suetonius T.G. 1st cent. AD. De vita Caesarum, liber V "Divus Claudius", passim 20-21.
Tacitus, P.C. 1st cent. AD. Annales XII, 56-57.
Titus Livius 27–9 BC., Ab urbe condita, liber V, caput III, 15,16,17,19.

Tunnels and Underground Cities: Engineering and Innovation meet Archaeology, Architecture and Art, Volume 1: Archeology, Architecture and Art in underground construction – Peila, Viggiani & Celestino (Eds)
© 2020 Taylor & Francis Group, London, ISBN 978-0-367-46574-2

Urban tunnelling under archaeological findings in Naples (Italy) with ground freezing and grouting techniques

F. Cavuoto
Project Manager of Metro Line 1, Naples, Italy

V. Manassero
Underground Consulting S.a.s., Pavia, Italy

G. Russo
University of Naples Federico II, Naples, Italy

A. Corbo
Geotechnical Engineer, formerly Studio Cavuoto, Naples, Italy

ABSTRACT: The paper deals with the design of twin tunnels, which construction has been completed on 2017, connecting the existing Municipio Station Line 6 main shaft (interchange node between Metro Line 6 and Metro Line 1 extension project in Naples, Italy) to the TBM extraction shaft located nearly 40 m far from it. The two short tunnels underpass archaeological findings consisting in fortification walls built in the sixteenth century, through a subsoil composed by loose silty sand overlying a soft and sometimes fractured rock (Neapolitan Yellow Tuff) and below the groundwater table. The design and construction processes have been very complex, adopting a mix of different measures as cement and chemical grouting at the sides and inverts in tuff and Artificial Ground Freezing at the crowns in sand, together with the compensation grouting technique, even for the obvious needs to preserve a unique archaeological site, part of mankind's common heritage.

1 INTRODUCTION

Municipio Station is located in the urbanized historical centre of Naples (Italy), facing the impressive backdrop of *Maschio Angioino* Castel and next to the tourist harbor *Molo Beverello*, and it will provide an efficient interchange node between Line 1 and Line 6 of the metro system serving the city (Figure 1). In particular, the new stretch of Linc 1 consists of five new stations and twin rail tunnels EPB-TBM excavated, with a total length of about 5 km, while Line 6, partially constructed several years ago, has been further developed with a 6 km new stretch with the use of mechanized excavation, and four new stations are under construction along it.

Currently, Municipio Station on Line 1 extension is open to the public, whereas Line 6 interchange portion has to be completed. Two tunnels, which construction has been accomplished on 2017, connect the existing Municipio Station Line 6 main shaft to the TBM extraction shaft, located nearly 40 m far from it.

During the works, important interferences between the excavations and the buried archaeological remnants occurred, unearthing part of the structures of the Roman port of the city and fortification walls built in the sixteenth century under the Spanish Viceroy, which, together with archaeological knowledge gained during the excavation works of the rest of the line, represent one of the most important archaeological sites in Europe (Giampaola 2009).

Due to the archaeological findings, the original design had to be changed to include part of the remnants into the new structures. For such reason, to construct the two short tunnels

Figure 1. Municipio Station schematic plan view with highlighted (in red) Line 1 and (in blue) Line 6.

connecting the access shaft to the TBM extraction shaft and to preserve the ancient structures subjected to constraint by the current regulations, a modification of the original tunnel design was proposed adopting a mix of different technologies to support the excavation: cement and chemical grouting with the Multi Packer Sleeved Pipe (MPSP) method, Artificial Ground Freezing (AGF) and the compensation grouting technique.

2 THE ORIGINAL DESIGN

The distance between the diaphragm walls of the previously excavated Municipio main shaft and the TBM extraction shaft is about 40 meters and the two connecting tunnels, slightly diverging in plan, have a polycentric cross-section about 6 meters wide and 7 meters high. The ground surface is between +12.25 and +3.00 m a.s.l. while the tunnel invert is located at an average elevation of -13.00 m a.s.l..

The subsoil is made by a loose silty sand of volcanic origins (*Pozzolana*) down to a depth between -11.85 and -7.25 m a.s.l., overlying a soft rock (Neapolitan Yellow Tuff) characterized by the presence of randomly distributed sub-vertical fractures (locally named *scarpine*), while the groundwater table is nearly horizontal in the whole area and located about 16 m above the invert of the tunnels.

To allow a safe and dry excavation via dewatering, a waterproof barrier with three lines of vertical jet-grouting columns (inserted into the tuff bedrock for a length of almost 1 meter) was executed to confine the volume between the two shafts (Figure 2).

Therefore, preceded by the lowering of the groundwater level through deep wells as per design requirements, the full section excavation had to be carried out under the protection of injected forepoling at the crown and along the upper part of the vertical sides into the granular soil, using fiberglass tubes to stabilize the front face and advancing with short steps of 1 m immediately followed by the installation of temporary steel ribs and shotcrete. At this point, an r.c. lining 0.6 m thick had to be cast in place along the whole section of the tunnel separated from the temporary one by a waterproof membrane.

2.1 *Settlements empirical prevision*

Settlement caused by tunnelling are mainly due to two components: the stress changes around the cavity due to soil removal and extraction, and the small excess of volume extraction to leave enough room to fit inside the excavated tunnel the temporary and permanent lining.

These movements are generally summarized and incorporated into the volume loss parameter, V_l (percentage of the volume of the full circular section with an equivalent diameter D):

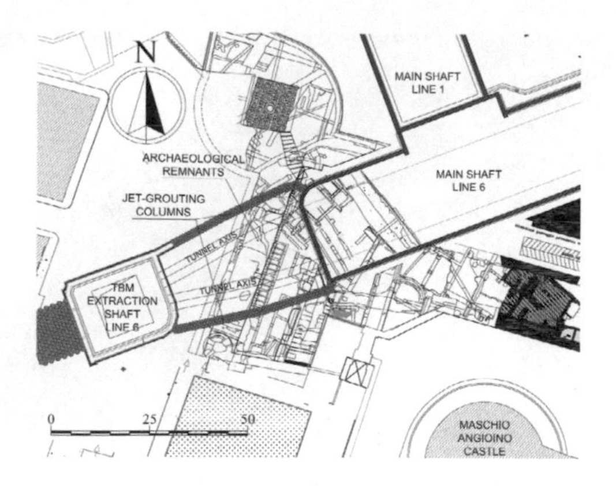

Figure 2. The two connecting tunnels below the ancient remnants, with the lateral jet-grouting barriers.

$$V_l = \frac{4V_s}{\pi D^2}, \tag{1}$$

where, V_s is the theoretical volume of the settlement trough per unit length of excavation which in the hypothesis of a Gaussian distribution curve (Peck 1969) is equal to:

$$V_s = i \times W_{\max}(2\pi)^{0.5}, \tag{2}$$

with i corresponding to the distance of the inflection point from the tunnel centre line and W_{max} being the maximum settlement occurring over the tunnel axis. On the basis of field data, O'Reilly & New (1982) suggested i be estimated as:

$$i = K \times Z_0, \tag{3}$$

where Z_o is the depth of the tunnel axis and K is a dimensionless trough width parameter.

By adopting values of the volume loss V_l within the limits of previously published ranges, as the ones reported by Bilotta et al. (2002, 2006) on the basis of a very large experimental database, maximum settlement obtained via the Equations 1 and 2 is shown in Table 1.

Before further commenting the obtained values, it is to underline that the adopted value of the inflection point distance i (i.e. the K parameter) is obtained by previous field data recorded during the tunnel excavation along the initial stretch of Metro Line 1 in Naples, carried out by both TBM and by conventional methods (Russo et al. 2015). According to the original suggestion by Peck (1969) the range of the measurements falls in the area which is intermediate between *sands above the groundwater table* and *sands below the groundwater table*. This result is consistent with the intermediate nature of the tunnel cover in the mentioned case histories.

The obtained Gaussian curves for the twin tunnels analysed separately are reported in Figure 3, regarding the smallest centre-to-centre spacing between the pairs of tunnels. In the same figure, assuming that the construction of the second tunnel is unaffected by the presence of the first one, the final curve settlement is reported by superposition of the two single tunnel predictions.

Table 1. Gaussian curves parameters for the settlement prediction of each tunnel.

V_l	D m	K	Z_0 M	W_{max} mm
0.01	5.2	0.52	20.4	7.8

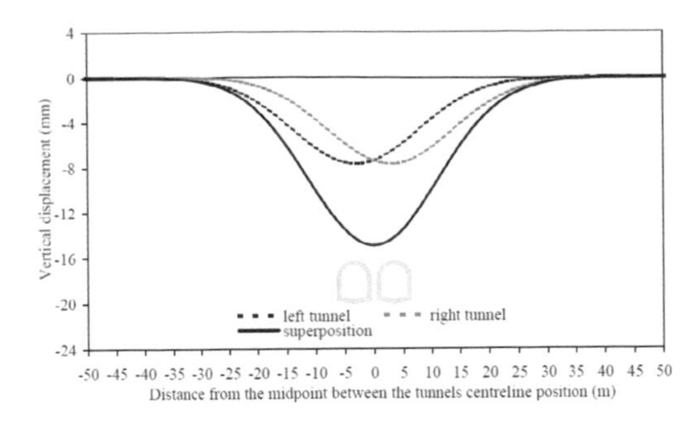

Figure 3. Calculated Gaussian settlement troughs and their superposition.

It is evident from Figure 3 that the contribution of the tunnel excavations in terms of induced displacements led to a total settlement of nearly 2 cm, not compatible with the safe preservation of the above archaeological site and its remnants. Besides, the controlled dewatering to permit the dry excavation will surely led to some additional settlement related at least to the increase in effective stress induced by the water level drawdown. This amount could be simply and roughly estimated by the formula proposed by Skempton & Bjerrum (1957) around 0.5 cm. This value has to be added to the previously calculated settlement induced by the stress relaxation due to the subsequent excavation, obtaining a total predicted settlement of more than 2 cm, which represented a potential cause of damage for the highly vulnerable archaeological findings.

2.2 *Dewatering field tests*

In order to verify the amount of settlement induced by the controlled lowering of the water table during the tunnels construction (and, therefore, the possible effects on the ancient structures) and in order to check the effectiveness of the jet-grouting barrier as hydraulic cut-off, a dewatering test programme has been conducted via no. 4 deep wells and a control system consisting in no. 13 piezometers installed both inside and outside the area of interest enclosed by the diaphragm walls of the two shafts and laterally by the jet-grouting columns, as shown in plan in Figure 4, together with the four wells location.

Furthermore, the monitoring program was setup and actuated relying mainly upon the survey on the surrounding buildings equipped with fixed benchmarks to evaluate the safety conditions in terms of induced settlement during the dewatering test.

During the test, as a result of the depression of the water table inside the confined volume, a significant groundwater level variation has been monitored also outside the soil bounded by the jet-grouting barriers, determining settlement at the foundation level of the surrounding

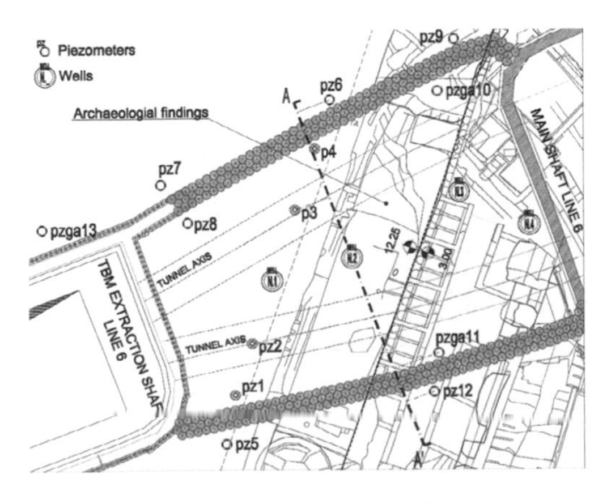

Figure 4. Plan view with the installed piezometers and the wells.

buildings. Therefore, the test was immediately suspended in order to avoid the development of crack patterns in the buildings masonry walls. This water level variation was likely associated with a water flow, induced by the pumping test, from the external area of the confined volume via unknown discontinuities (*scarpine*) in the tuff formation, resulting as hydraulic connection between the confined volume to be dewatered and the surrounding soil.

Consequently, as evidenced by the precision leveling on the surface, settlements appeared and were judged as not compatible with the safeguard of the existing adjacent buildings and the overlying ancient structures. Besides, the pumping test carried out led to unsatisfactory results for the execution of the excavation of the tunnels as defined at the design stage, because despite the pumped high flow (overall about 37 l/s) the water table could not be lowered below the tunnels invert location, a necessary condition to allow the dry excavation of the two tunnels with conventional methods.

3 MODIFICATION OF THE ORIGINAL TUNNELLING TECHNIQUE

3.1 *AGF and MPSP: adoption of a mixed approach to support the excavation*

As evidenced by the data of the above mentioned topographic survey, the lowering of the groundwater also outside of the volume confined by the jet-grouting barrier has determined both average and differential settlement on the adjacent buildings and on the archaeological findings. Such sagging, even if of the order of millimeters, was considered as significant warning signal comparing the small duration of the pumping test (2 days) with the likely duration of the excavation works, and then with the period in which it would have been necessary to lower the water table (i.e. 3-4 months).

Therefore, the final design solution was accordingly modified to permit a safe and substantially dry excavation without lowering the groundwater level via pumps and preserving both the above archaeological findings consisting in fortification walls of the sixteenth century and the modern buildings next to the working area.

A mix of different measures has been combined in the design: cement and chemical grouting at the lower part of the sides and at the inverts (in tuff) and Artificial Ground Freezing at the crowns and at the upper part of the sides (in sand). The compensation grouting technique was also included to actively protect the historic tuff walls from damages during the construction stages. In Figure 5, a sketch of the cross section with the soil treatments to support the surrounding soil during the excavation of the twin tunnels is reported, designed in function of the variation level of the roof of the Yellow Tuff formation.

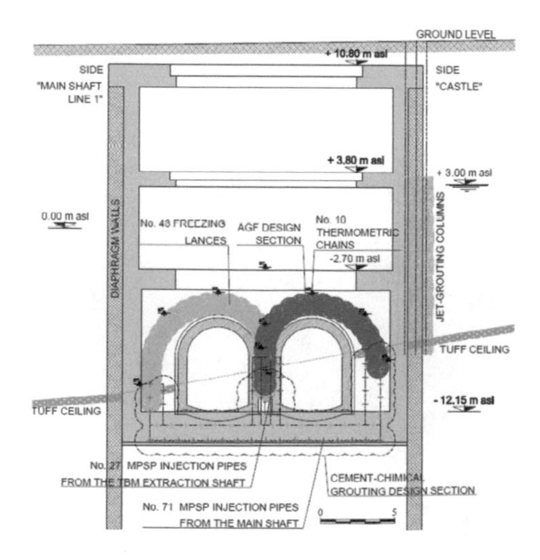

Figure 5. Schematic cross section of the two tunnels and soil treatments adopted.

Due to construction reasons, boreholes for the soil treatments carried out in presence of Blow Out Preventer (BOP), were conducted from both shafts. In particular, no. 27 horizontal boreholes for grouting at the upper sidewalls and no. 43 directional drillings at the crown for the installation of freezing pipes were carried out from the TBM extraction shaft, while, no. 71 horizontal boreholes for grouting at the lower sidewalls and at the invert together with no. 27 directional drillings for compensation grouting were instead performed starting from the main shaft Line 6.

The grouting method adopted was the Multi Packer Sleeved Pipe. The system consisted essentially in the installation inside totally no. 98 boreholes of a plastic or steel pipe equipped at regular intervals with rubber grouting sleeves and bag packers fastened to the grouting pipe (expanded against the hole walls through grout injection into the bags) to seal off the sections to be grouted. The main role of the grouting is to improve the mechanical properties and to reduce the hydraulic permeability of the fractured rock. Initially, cement grouting are used to fill the thicker cracks; afterwards chemical grouting have to fill the partially treated material, permeating then the remaining thinner cracks (Manassero & Di Salvo 2015).

The AGF technique was already extensively adopted in the construction of Line 1 of Naples underground during tunnels excavation below the groundwater table (Manassero et al. 2008, Viggiani & De Sanctis 2009, Russo et al. 2012, Cavuoto et al. 2015, Viggiani & Casini 2015, Russo et al. 2017). The application is based on withdrawing heat from the soil thus converting in-situ pore water into ice; the ice binds the soil particles imparting strength and very low permeability to the frozen soil mass. For the two tunnels AGF was carried out through no. 43 horizontal freeze lances installed from the main shaft, disposed along a single row and drilled by using the Horizontal Directional Drilling (HDD) technique.

AGF was thus adopted to create a frozen arch around each ceiling of the twin tunnels to be excavated in the sandy soil, in order to temporarily ensure stability and waterproofing of the tunnels crown during the construction process until the final lining was installed. The construction designed sequence started with the left tunnel, (side named “main shaft Line 1”) activating the ground freezing system first and excavating the tunnel, finally installing the supporting liners. The refrigeration plan was then moved to work for the right tunnel construction following the same steps.

During the freezing phase, the initial use of Liquid Nitrogen as coolant medium (entering at -196 °C into the freeze tubes) allowed to form the ice arch very quickly. After a fast lowering of the temperature down to peaks of about -60 °C the frozen arch was completed achieving the design thickness of about 1.5 m. The temperature was then kept stable between -15°C and -25°C for the time needed to excavate the full section of each tunnel, by using the less expensive brine (a calcium chloride solution in water) as coolant fluid chilled and re-circulated by a refrigerating plant, until the final lining has been constructed.

3.2 *Finite Element analyses*

At the design phase, early analyses have been conducted to simulate the freezing effects, the tunnels excavation and the thawing response, the prediction of the increase of the settlement at the end of the thawing process being the target quantity, assigning increased strength and stiffness to the frozen arch during freezing, and then, reduced mechanical parameters to simulate the usual soil deformation after thawing.

The analyses have been implemented in a 2D FEM model taking into account the main steps of the full construction process. Initially, the ground freezing of the left tunnel crown has been introduced through the sudden change of the mechanical properties of the treated material and also by not allowing any seepage flow.

Then, the analysis proceeded by the left tunnel excavation with the soil removal inside the design section of the tunnel itself (with typical stress reduction methods) and the subsequent installation of the first and second liner as plate structural elements. The thawing stage was the last step of the analysis, which was simulated assigning reduced mechanical properties of the soil volume involved by AGF. After this, the same phases sequence has been applied to model the construction of the right tunnel.

Though more advanced constitutive models could be more appropriate to better calculate the induced settlement field, the simple Mohr-Coulomb model with a linearly elastic initial part was preferred to focus on the capacity to predict the maximum settlement, via a simple imposed degradation of the stiffness and of the strength of the frozen and then thawed materials, as confirmed by different studies regarding preliminary experimental investigation in cryogenic conditions on the behavior of pyroclastic soils and available field data on the adoption of AGF in the Neapolitan subsoil (Pelàez et al. 2014, Russo et al. 2015, Casini et al. 2016).

In Table 2 the main physical and mechanical properties of the soil layers before and after the treatments are reported. In Figure 6 the contours of the calculated settlements are reported corresponding just to the final step of the construction process, regarding the section adjacent to that one (no.1-1) sketched in Figure 5 and located in the layout in Figure 7, characterized by the maximum cover and the minimum horizontal span between the two tunnel axes.

It can be seen that even if the calculated trough was confined by the previously made structural elements (jet-grouting and diaphragm walls), preserving the adjacent buildings from displacements produced by the excavation, the simulated thawing process inducing degradation of strength and stiffness of the soil after a freeze-thaw cycle, caused settlement not consistent with the preservation of the above valuable remnants. For this reason and taking into account also the uncertainties related to such complex techniques, additional suitable mitigation techniques were adopted during the construction stages to compensate the possible settlements due to the construction works.

3.3 *Compensation grouting and monitoring programme*

During the extensive and successful application of AGF to construct many tunnels for platforms and pedestrian passageways along Metro Line 1 stations, ground subsidence has been

Table 2. Soil parameter adopted.

		Sand	Tuff	Frozen sand	Injected tuff	Thawed sand
γ_d	[kN/m^3]	15	14	17	16	15
γ_{sat}	[kN/m^3]	17	16	17	16	17
E	[MPa]	150	1500	300	1500	100
c	[kN/m^2]	-	1000	160	1000	-
φ'	[°]	35	28	35	28	26
ν	[-]	0.3	0.3	0.3	0.3	0.3
K_0	[-]	0.426	0.531	-	-	-
Perm.	[m/s]	$1.00{\cdot}10^{-5}$	$5.9{\cdot}10^{-6}$	-	-	$1.00{\cdot}10^{-5}$

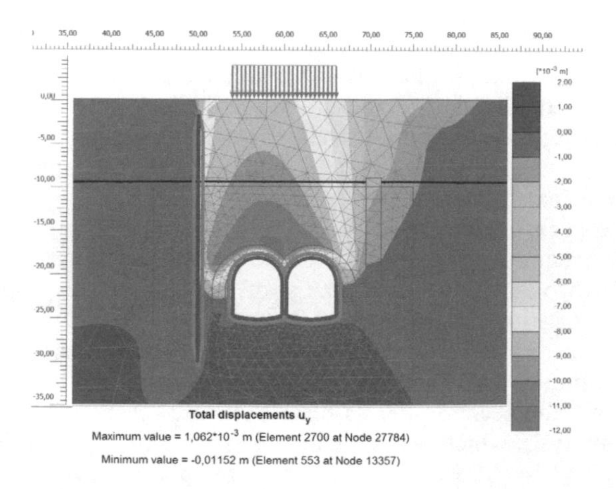

Figure 6. Contour map of the calculated settlement related to the final stage of the simulation.

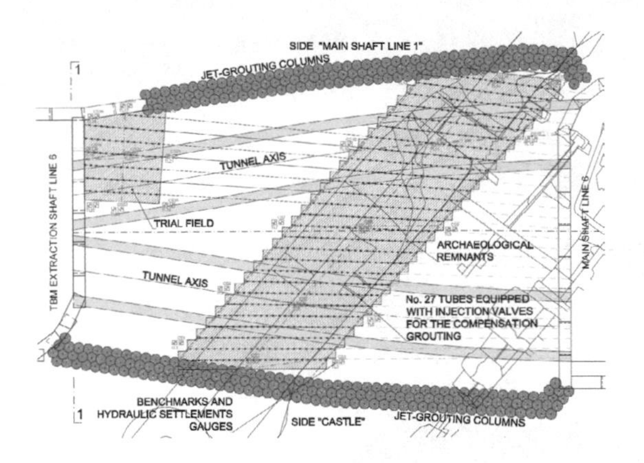

Figure 7. Detail of the designated area for the compensation grouting with TAMs.

systematically observed via the intense programme of monitoring, whose settlement has been correlated not only to the usual stress relaxation and volume change related to the tunnel excavation but also to the complex freeze-thaw processes of the involved pyroclastic soils.

In fact, during freezing the soil experiments a volume expansion associated with the change of pore water to ice. On thawing, the ice disappears and the soil skeleton must adapt itself to a new equilibrium void ratio according to the existing overburden pressure: drainage of the thawed soil and soil structural changes that occur during the previous freezing operation lead usually to additional volume change (Andersland & Ladanyi 2004), the amount depending on the extent of the soil portion involved in AGF.

For this reason, surface frost heave (usually negligible in granular soils) during the freezing stage and subsidence phenomena induced by thawing when the refrigeration system is de-activated, are substantially characteristic of this technique.

In order to minimize any effect on the archaeological remnants regarding the tunnels excavation and the adoption of AGF technology, the execution of the twin tunnels was supported by suitable grout injections during the construction phases and in particular, during the thawing process. The location of the tubes *à manchettes* (TAMs) and their portion equipped with injection valves is represented in Figure 7.

Compensation grouting is in fact an injection technique used to prevent or limit as much as possible subsidence induced by underground works, avoiding possible damages on the structures that insist on the subsidence basin. It is a technique called “active”, based on coordination and synchronization between monitored settlements and injections. Real time data allow to control and immediately correct the movements induced by the underground constructions.

The operating principle of compensation grouting is based, in this case, on the injection of cement mixture through tubes *à manchettes*, in the portion of soil between the two tunnels and the foundations of the historical walls, injection that took place in particular during the thawing phase of each ceiling tunnel previously frozen.

Below the archaeological finds, then, no. 27 horizontal drilling have been provided for a total length of 742 m of pipes, about 1.5 m horizontally spaced and 1.5 m vertically spaced. The holes had to be drilled below the water table in the loose material, placed 0,60 m above the deck at a height of -2.98 m a.s.l. of the main shaft.

Below the groundwater level all the drillings and the subsequent installation of the pipes were carried out with the adoption of the Blow Out Preventer. The no. 27 tubes, installed via the Horizontal Directional Drilling technology, had a total valved length equal to 355 m (no. 709 valves at intervals of 0.50 m).

The injections were carried out in a repeated and selective way with suitably studied mixtures, in correspondence of the footprint of the wall foundations, as reported in the layout in Figure 7, during the two thawing stages of the soil at the crown of each tunnel subject to AGF.

Control of vertical movements, injection volumes and pressures in real time were possible due to the use of a specific monitoring program adapted to keep automatically and

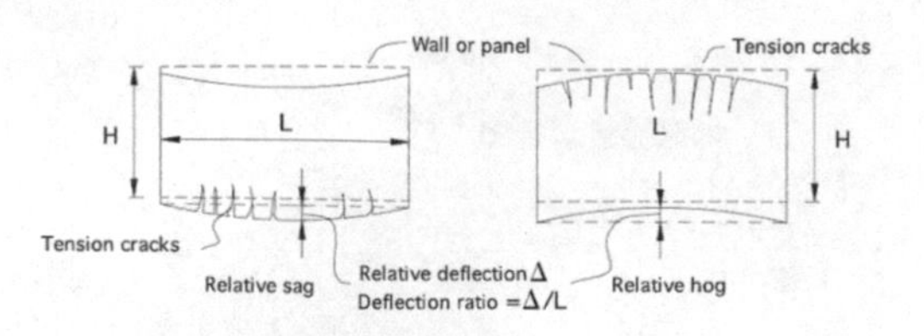

Figure 8. Scheme of possible foundation deformations.

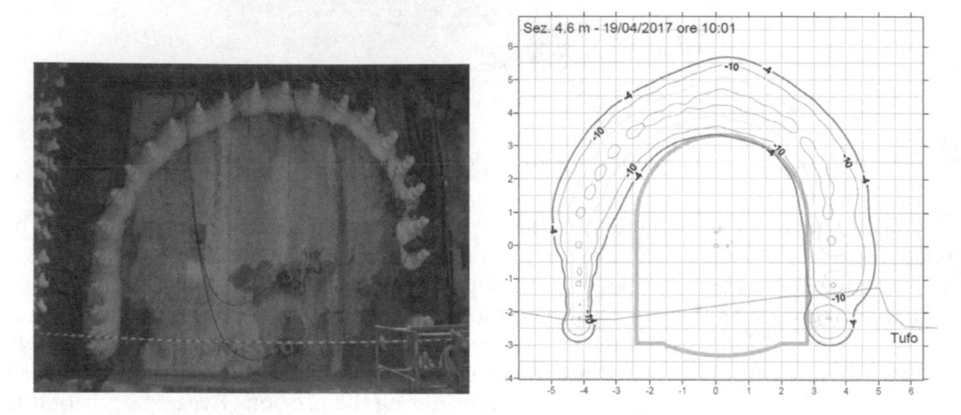

Figure 9. On the left, freeze and temperature pipes around the tunnel section; on the right, the isotherm contour map of the frozen body.

continuously measurements during the phases of injection. In the already mentioned Figure 7 also the position of the integrated topographical monitoring equipment (no. 15 coupled benchmarks and no. 15 hydraulic settlements gauges) with the imprint of the below tunnels and the previous consolidation works are sketched.

In the scientific literature (Viggiani et al. 2011) limit values of the deflection ratio, as the relative deflection D divided by the distance between two points L (Figure 8), are reported beyond which there are possible manifestations of damages in case of masonry structures. These values, equal to 1/1250 in sagging and 1/2500 in hogging were also prudently referred to the ancient findings object of intervention, in the design phase. During works, measured deflections, when close to these limits, represented an effective threshold level to start the corresponding compensation injections, avoiding the occurrence of damages to the existing masonry ancient structures.

Besides, no. 10 sub-horizontal thermometric chains (instrumented by thermocouples arranged at intervals of 5 m) have been installed to obtain real-time distribution of temperatures in the frozen ground during freezing, maintenance and thawing phases and to have direct information on the continuity and efficiency of the frozen arch, with particular reference to the isotherm -2°C, which is usually assumed to be the isotherm limit of the frozen body (Figure 9).

4 CONCLUSIONS

The twin tunnels, connecting the existing Municipio Station Line 6 main shaft (interchange node between Metro Line 6 and Metro Line 1 extension project in Naples, Italy) to the TBM extraction shaft were located under a valuable archaeological site where remnants dating back to the sixteenth century were found. The original design had to be modified in order to permit the safe tunnel excavation excluding any even minor negative effect on the ancient structures. So that, different soil treatments have been designed.

In particular, cement and chemical grouting at the lower part of the sides and at the inverts (in tuff) and the Artificial Ground Freezing at the upper part of the sides and at the crown (in sand) were employed. The indication coming from the applications of AGF technique on Line 1 in recent years was that in volcanic soils, such as *Pozzolana* and tuff, rather large settlements may be induced during the thawing stage. In this project the compensation grouting technique was included in order to successfully preserve the historic fortification walls built in the sixteenth century from potential damages during the complex construction stages, because of the typical ground subsidence produced by both the stress relaxation and volume change related to the tunnel excavation and by the freeze-thaw cycle related to the adoption of AGF.

REFERENCES

Andersland, O.B. & Ladanyi, B. 2004. *Frozen ground engineering*, Second Ed., New Jersey: John Wiley & Sons.

Bilotta, E., Russo, G. & Viggiani, C. 2002. Cedimenti indotti da gallerie superficiali in ambiente urbano. *Atti XXI Convegno Nazionale di Geotecnica*, Bologna: AGI.

Bilotta, E., Russo, G. & Viggiani, C. 2006. Ground movements and strains in the lining of a tunnel in cohesionless soil. *Proc. 5th Int. Symp. Geotech. Aspects of Underground Construction in Soft Ground*: 705-710. Leiden: Taylor & Francis/Balkema.

Cavuoto, F., Corbo, A., Manassero, V. & Russo, G. 2015. Naples Metro Line 1: the service tunnel at Toledo Station, *Gallerie e Grandi Opere Sotterranee, n.116, December 2015*: 9-19. Bologna: Pàtron Editore.

Casini, F., Gens, A., Olivella, S. & Viggiani, G.M.B. 2016. Artificial ground freezing of a volcanic ash: laboratory tests and modeling. *Environmental Geotechnics*, 3(3): 141-154. London: ICE Publishing. (http://dx.doi:10.1680/envgeo.14.00004).

Giampaola, D. 2009 Archeologia e città: la ricostruzione della linea di costa. *TeMA – Trimestrale del Laboratorio Territorio Mobilità Ambiente*. 2(3): 37–46. Napoli.

Manassero, V., Di Salvo, G., Giannelli, F. & Colombo, G. 2008. A combination of artificial ground freezing and grouting for the excavation of a large size tunnel below groundwater. *Proceedings of the 6th International Conference on Case Histories in Geotechnical Engineering*, Arlington (VA).

Manassero, V. & Di Salvo, G. 2015. The application of grouting technique to volcanic rocks and soils to solve two difficult tunnelling problems. In *Proc. of The International Workshop on Volcanic Rocks and Soils*: 399-408. Ischia Island: AGI.

O'Reilly, M.P. & New, B.M. 1982. Settlements above tunnels in the United Kingdom - Their magnitude and prediction. *Proceedings of Tunnelling'82 Symposium*: 173-181. London: Institution of Mining and Metallurgy.

Peck, R. B. 1969. Deep Excavations and Tunnels in Soft Ground. *Proceedings of the 7th International Conference on Soil Mechanics and Foundation Engineering, State of the Art Volume*: 225-290. Mexico City: Sociedad Mexicana de Mecanica.

Pelàez, R.R., Casini, F., Romero, E., Gens, A. & Viggiani, G.M.B. 2014. Freezing-thawing tests on natural pyroclastic samples *6th Int. Conf. Unsaturated Soils, UNSAT2014*: 1689-1694.

Russo, G., Viggiani, C. & Viggiani, G.M.B. 2012. Geotechnical design and construction issues for lines 1 and 6 of the Naples underground. *Geomechanik und Tunnelbau* 5(3): 300–311. Berlin: Ernst & Sohn.

Russo, G., Corbo, A., Cavuoto, F. & Autuori, S. 2015. Artificial Ground Freezing to excavate a tunnel in sandy soil below groundwater table. Measurements and back analysis. *Tunnelling and Underground Space Technology*, 50: 226-238. The Netherlands: Elsevier. (doi:10.1016/j.tust.2015.07.008).

Russo, G., Corbo, A., Cavuoto, F., Manassero, V., De Risi, A. & Pigorini, A. 2017. Underground culture: Toledo station in Naples, Italy. *Proceedings of the Institution of Civil Engineers - Civil Engineering, November 2017*. 170(4): 161-168. London: ICE Publishing. (doi:10.1680/jcien.16.00027).

Skempton, A.W. & Bjerrum, L. 1957. A Contribution to the Settlement Analysis of Foundations on Clay, *Geotechnique*, 7: 168. London: ICE Publishing.

Viggiani, C., Mandolini, A. & Russo, G. 2011. *Piles and Pile Foundations*. 278 pp. London: CRC Press, Taylor & Francis.

Viggiani, G.M.B. & De Sanctis, L. 2009. Geotechnical aspects of underground railway construction in the urban environment: the examples of Rome and Naples. *Geological Society Engineering Geology Special Publication*. 22(1). 215–240. London: The Geological Society

Viggiani, G.M.B. & Casini, F. 2015. Artificial Ground Freezing: from applications and case studies to fundamental research. *Invited Papers*: 65-92. London: ICE Publishing. (http://dx.doi:10.1680/ecsmge.60678.vol1.004).

Tunnels and Underground Cities: Engineering and Innovation meet Archaeology, Architecture and Art, Volume 1: Archeology, Architecture and Art in underground construction – Peila, Viggiani & Celestino (Eds)

Integration of archaeology in architectural design of Milan Metro connection M2–M4 in St. Ambrogio Station

M.N. Colombo, A. Bortolussi & E. Noce
MM Spa, Milan, Italy

ABSTRACT: The central section of new metro line M4 in Milan cross an high-density urban area with a lot of issues cause by historical and archaeological building, one of the most complex case is the St. Ambrogio station. The initial solution for the connection between existing metro line M2 and the new one M4 in St. Ambrogio station was to build a 100 m long rectangular open-air tunnel placed above and practically on the same axis with the existing tunnel of M2 built in the 80s with a traditional excavation method, with a gap between the two facilities being quite strait, of about 1.50 m. Such a solution created problems for the further development of the overall project in terms of the traffic/construction site organization, because the connection had to be placed exactly along an important traffic corridor making part of internal ring road of the city of Milan. An alternative engineering and architectural solution has been developed in collaboration with the archeological supervision authorities. The integration and utilization, in the tunnel connecting M2 and M4, of the belowground part of the complex of Pusterla of St. Ambrogio, built in XII century a.C. and under protection of the archeological supervision authorities, helped to avoid a complete surface traffic block, to resolve a number of critical issues from the geotechnical and structural points of view, and to revaluate the archeological site that was incorporated into the connection between the M2 and M4 metro stations.

1 GENERAL INFORMATION OF NEW METRO LINE M4

The construction of Line 4 of the city of Milan has been set up and managed in a Public Private Partnership (PPP); for this purpose an ad hoc company has been set up, two-thirds owned by the Municipality of Milan and a third by the private partners in charge of designing, constructing and managing the new line 4 of the Milan underground (M4 Spa society).

The duration of the concession is approximately 31 years (370 months) from the date of signature of the Convention (December 22, 2014), of which 88 months for the construction of the entire work.

The Private Members, selected through a specific call for tenders, are formed the CMM4 consortium consisting in a temporary grouping of the companies Salini-Impregilo, Astaldi, Ansaldo STS, AnsaldoBreda (later Hitachi Rail), ATM and Sirti. The same is given the burden of design the Executive Project of the work, as well as the realization of the same (EPC contract). Within the above framework, MM company, in addition to drawing up the project in the preliminary phase, has had a controlling role in the executive design developments, the Technical Assistance to the Client, the Safety Coordination and Work Supervisor.

The new metropolitan infrastructure create a fast public connection along the east/south-west route through the historic city center, developing for 15km from San Cristoforo to Linate, with 21 stations, 30 line shafts and 1 depot-workshop.

It is an automatic metro without driver on board (driverless) with automatic dock doors and CBTC signaling system (Communication Based Train Control); the fleet consist of 47 vehicles.

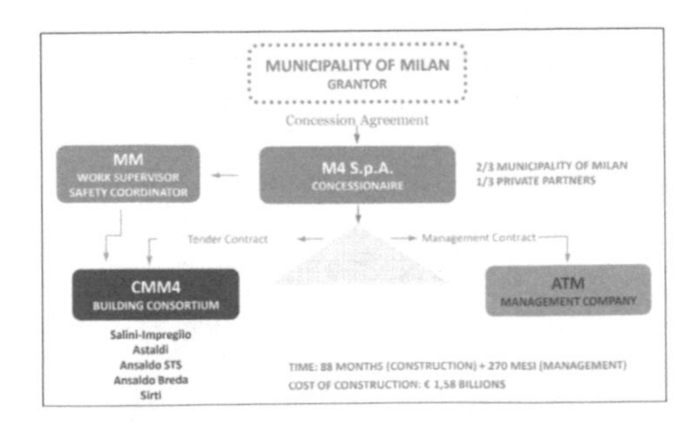

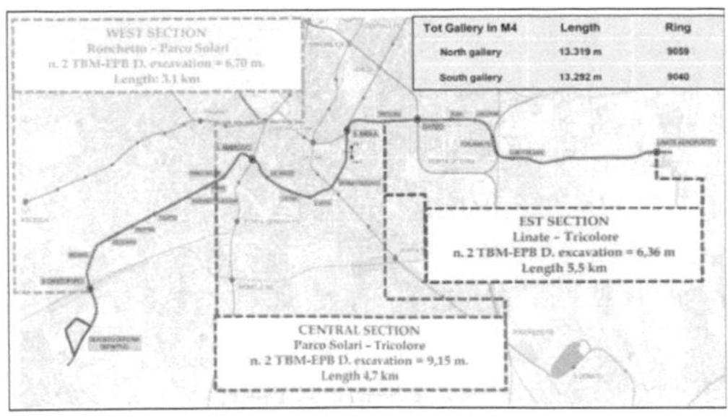

Figure 1. Contract scheme and general planimetry of line M4.

There are two interchanges with the existing metro lines, one with the M1 line, at the San Babila station, and one with the M2 line (green), corresponding to the St. Ambrogio station. A further connection with the yellow line (M3) is being studied at the Porta Romana course.

There be three interchanges with the suburban railway lines at Forlanini FS station, Dateo station and San Cristoforo station, and one with Linate airport.

Most of the underground development of the track is made by mechanized excavation, using two TBM geometries: one with an excavation diameter of 9.15 m, in the central section to accommodate the station platform directly in the gallery; the other with a smaller excavation diameter (D = 6.36/6.70 m), for the realization of the peripheral sections. In total, be used n. 6 TBM EPB for tunnel excavation.

2 ARCHEOLOGY IN M4

2.1 *General overview of archeology in M4*

The goal of bridging the infrastructure gap between the south-west and east areas of the city (compared to the center), of certain usefulness and ambition, has generated a railway track that has considerably involved a part of the center of Milan, although aware that this would have necessarily involved a series of inconveniences to the city and to the citizens. Among these, of sure importance are the management of vehicular and pedestrian traffic, the management of relationships with the existing and even more commercial realities, but also the interface of the construction sites with a multitude of pre-existences with a strong historical and architectural interest and/or archaeological.

On the other hand, Milan was founded about 2,600 years ago, beginning an ascent that led it to be a cultural, commercial and military center among the most prosperous in Northern Italy since Roman times and, on several occasions, also Capital of the Roman Empire West. A thousand-year history of such importance can not but leave traces of itself and in fact even Milan does not escape this rule of Italian population centers. There are in fact innumerable portions of cities that have returned, over time, more or less important signs of the past epochs. The above map shows the pre-existences known and considered relevant from a historical, archaeological, architectural and monumental point of view. Naturally the greatest denseness of findings regards the city center; to name a few are the amphitheater, the circus, the Erculee thermal baths, the access gates to the fortified city, as well as a series of palaces from different historical periods each, in its own way, an important hallmark of its era. The overlapping of this map with the M4 layout makes clear how inevitable the theme (citing the Archaeological Superintendence of Milan) of the relationship between modern needs and protection of the past would have been inevitable.

The forecasts referred to this paragraph have actually been complied with, not without some surprises. Infact, in the course of the works, with particular reference to the sites located in the most central area of the city, numerous archaeological finds were found, among which Roman and medieval masonry structures, funeral objects, foundation of ancient monument Pusterla of Fabbri, Roman baths, burial grounds.

What has been found has become part of a specific project dedicated to works for Line 4 entitled "Time travel with metro line M4" on show for 3 month (September-December 2018) at the Archaeological Museum of Milan sponsored by the City of Milan, in collaboration with the Contractor, in order to spread the knowledge of Milan that was.

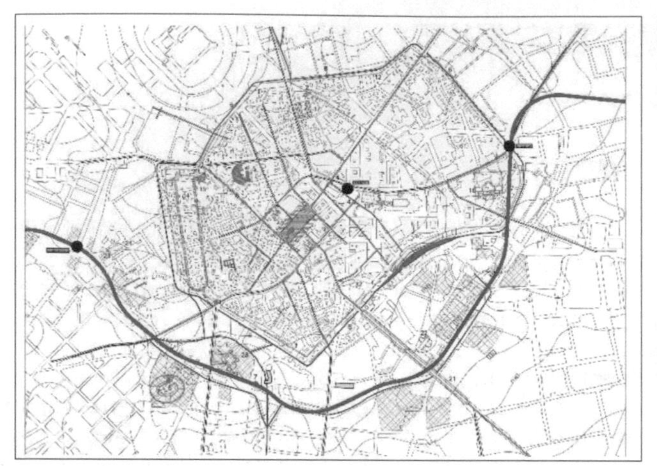

Figure 2. Plan of archaeological risk areas and archaeological finds.

2.2 *Archaeological procedures in M4*

Since 2004, thanks to the collaboration between the subjects interested in the realization of line 4 and the Italian authority responsible for the protection of cultural heritage (Ministry of Cultural Heritage and Tourism), a series of in situ verifications have been studied and agreed to give a series of design indications with respect to possible interference of the M4 with elements of archaeological importance.

The collaboration of all the Parties has also produced a specific agreement for the discipline of the archaeological theme, in which specific procedures are followed to assist specialized personnel during the construction of the work and any further intervention procedures in case of discovery finds of archaeological interest. In particular, for the central section of the Line, which is developed from the S. Vittore shaft to the S. Damiano shaft (included), there have been other cases.

First, the excavations are subdivided by type between:

1. excavations necessary for all the works as a corollary of the mere realization of the profund body, such as for example the sub-services to be modified to free the areas;
2. excavation of the bodies of work (stations and artifacts).

In the first case, considering the widespread reorganization that took place in the last decades and concerning the city's infrastructural networks, it was substantially taken into consideration the scarce probability of finding archaeological evidence, at least up to a depth of about 2m from the road level; from there and up to the share of sterile soil, tests were carried out with metal detectors for layers of one meter in order to intercept any suspicious masses.

Within the impression of the bodies of work the above-mentioned intervention logic has remained valid up to 2m depth from the countryside, then use the geo-radar for layers of two meters up to the share of the sterile soil and then authorize the second phase of research of war bombs (the deep one).

During all these phases the presence of specialized personnel is always guaranteed, operating according to two possible approaches:

- Assisted Excavation - consisting in the presence of qualified personnel during all excavation work until reaching the sterile layer of soil;
- Archaeological excavation - implemented when something significant has been found and which provides for the suspension of work to allow an excavation of an archaeological type with manual means adapted to the type of discovery that emerged.

3 ST. AMBROGIO STATION AND HIS CONNECTION WITH METRO LINE M2

3.1 *Archaeological-monumental & Architectural-structural overview*

The St. Ambrogio Station (ST09) is definitely a paradigm of the above. It is part of one of the most complex junctions of the city, grafted on two major roads such as Carducci street (north-south road axis) that connects the Cadorna railway station to the first ring road of Milan, and San Vittore street (east-west axis) along which overlook, within 300 meters, the San Giuseppe Hospital, the Basilica of San Vittore, the National Museum of Science and Technology, some historic buildings (including Cova Castle), some pre-existing archaeological sites including the remains of the Maximinianus Mausoleum and the Pusterla (ancient gateway to the medieval city), up to the Basilica of Sant'Ambrogio (one of the most important churches extended, as well as being the symbol of the city itself) and the Catholic University of the Sacred Heart (among the most important universities of Milan). To this is added the presence of the homonymous subway station Line 2 (green one) to which you have to connect, in addition to a multitude of public transport lines and private traffic enjoyment of the aforementioned streets and commercial realities to be preserved.

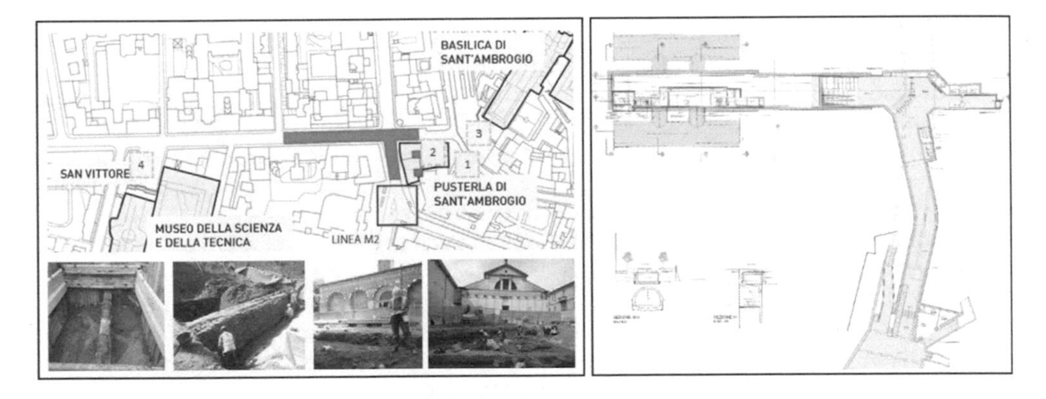

Figure 3. Plan of St. Ambrogio M4 station and archaeological surrounding area.

With great care and a good dose of ingenuity, it was possible to define a space sufficient for the creation of a typological subway station along the San Vittore street, in the section between Carducci and De Togni streets.

The St. Ambrogio station is one of the so-called deep stations of Line 4, those stations which, due to unfavorable boundary conditions, had to deepen more than the others to avoid hindering existing networks and structures. It is a station that engages between the two gallery (even and odd), connecting to them by a series of connecting tunnels, developed in only 13 meters of width within a road from the overall width (between the buildings) of about 16 meters.

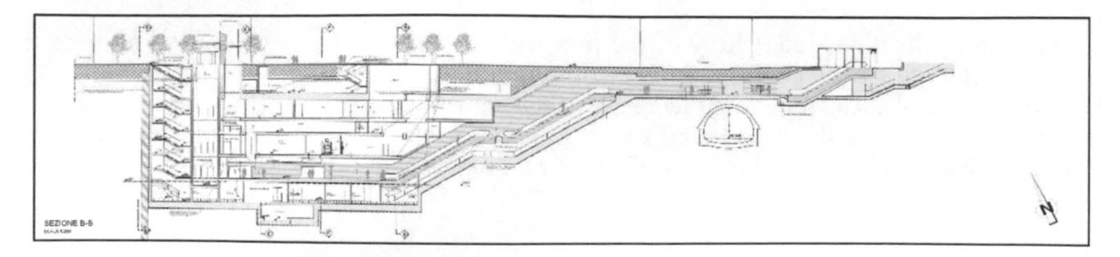

Figure 4. Longitudinal section of St. Ambrogio M4 station and the section of M2 gallery.

3.2 *Problems connected to the traffic node*

The complexity of the road network S. Vittore - Carducci streets, right from the first phase of the study of the site, appears one of the most critical knots of the entire M4 line, being via Carducci an integral part of the inner ring road of the city and transit center of transport lines local public. The works of the S. Ambrogio station, given the density of the urban fabric in which it is located, can only occupy the surface road surfaces.

Specifically, the works of the station body completely occupy the west axis of S. Vittore street, the corridor connecting the M2 line (rectangular box 7,20x3,6 m) extends south for about 100m, completely occupying Carducci street and the atrium floor with the turnstile access line of the station is located right in the central part of the road crossing.

Moreover, the complex execution of the works of the atrium and corridor connecting M2–M4, which are performed with excavation in open air, without the use of diaphragms or micropiles, being placed above the tunnel of the M2 underground line (frank between the two structures equal to 1.06 m), metro line in operation and built in the 80s with traditional excavation method. To mitigate the impact of the works, the station, according to what is foreseen by the definitive project, is carried out in four phases of construction that provide for the execution of the central body of the station and of the lift tunnels, before moving on to the construction some parts of the tunnel below the intersection between S. Vittore and Carducci street and the stairwell on the south-west side; in the third phase the tunnel along S. Vittore street is developed, in addition to the construction of the stairwell on the South-East side of the station, and finally the internal completion of the station.

In the final design configuration, in addition, the ascent stairs in line were oriented towards Carducci street, with arrival, near the intersection with S. Vittore street, at a mezzanine level useful to realize also a tunnel connecting to the homonymous underground station of Line 2. From the mezzanine floor it was foreseen the possibility of exit on S. Vittore and Carducci streets, both on the south side of the station. Then there was a further access by elevator in the west (near the S. Giuseppe Hospital) that connected directly to the quay floor and, therefore, available to people with motor disabilities.

However, this configuration left unresolved a series of problems, highlighted during the course of the project authorization processes, including:

- The total occupation of the crossing between the Carducci-S. Vittore streets for considerable time, with consequent important effects on citizenship;
- The need to interface with the pre-existing historical-archaeological-architectural relevance, in order to harmonize the project to the needs of protection of cultural heritage, with particular reference to the stairs and exit elevators;
- The occupation of the entire Carducci street for the construction of the connecting corridor towards M2 station;
- Management of unbalanced exit and entry flows towards the intersection of the Carducci-S. Vittore streets, with penalization of the eastern portion and concentration of the entire flow towards a single exit point.

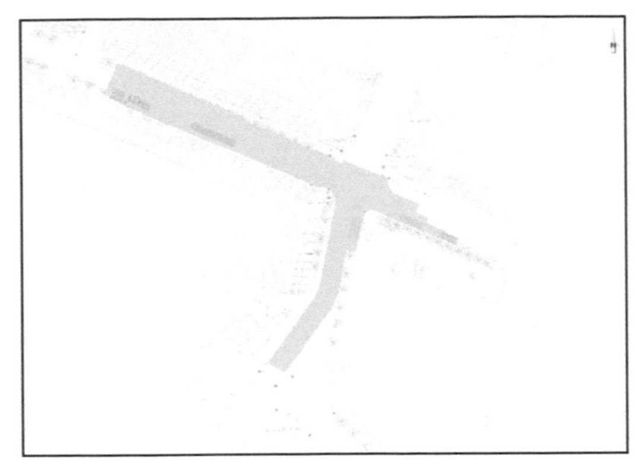

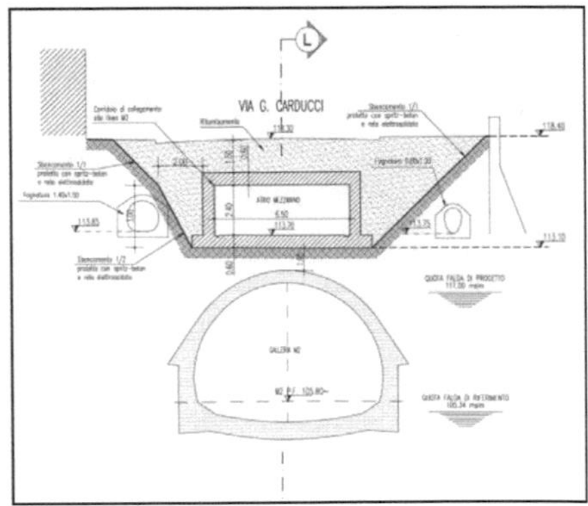

Figure 5. Traffic node S. Vittore-Carducci and tunnel section of M2–M4 connection.

4 Project Solution

4.1 *Change to the architectural-functional design of St. Ambrogio station*

The in-depth analysis of the aforementioned themes by all the parties involved, has generated a design review capable of resolving the critical issues and optimizing their implementation, both in terms of pure feasibility, and in terms of implementation.

First of all, it was decided to modify the construction sequence of the civil works, anticipating the construction of the roof slab and then working in top-down. The possibility of having already at the first stage of the roofing slab in S. Vittore street will make it possible to improve the pedestrian paths by moving away the building fences from the buildings and shops and reducing the superficial impact of the building site.

The compositional scheme maintains the structure of the station (as a shaft) between the two galleries, while the internal distribution has been completely modified. Further changes were introduced, that reduced the impact on the road network, such as the top down construction of the roof slab and the use of the provisional struts in place of the tie rods, to reduce interference with the surrounding buildings.

The essential feature of design variation is, however, represented by the crossed escalator scheme instead to the continuous drop-down scheme of tender design. Consequently, the longitudinal development of the stairway block is considerably reduced, since the spaces dedicated to the atrium and the platform floor are superimposed. This also implies a substantial difference in the distribution of the station spaces, with a clear separation between the service block and the block open to the public.

The new double-exit architectural and functional configuration has allowed the division into two of the turnstiles lines, decreasing the pedestrian flow volumes pertaining to M4 to be built in the stretch below the intersection Carducci-S. Vittore streets and so the atrium surfaces. In fact, in the definitive project just under Via Carducci the turnstile was located with a greater consequent area affected by the workings and a consequent greater closing time of Carducci street, with a strong impact on the road network.

Minimized the closing time of the Carducci/San Vittore intersection for the construction works of the station body, two design revisions were developed for the connection M2–M4:

- a first coherent with the definitive design tracing, which provided for the construction of the tunnel on the axis of Carducci street, up to the connection with the pre-existing station of metro line 2;
- a second one, which, instead, involves the translation of the connecting corridor towards the east side of via Carducci, realizing the accesses to S. Ambrogio square no longer on via di S. Vittore, but on the hypogeum part of the tower of Pusterla, which is at the same level with the entrance hall to the station.

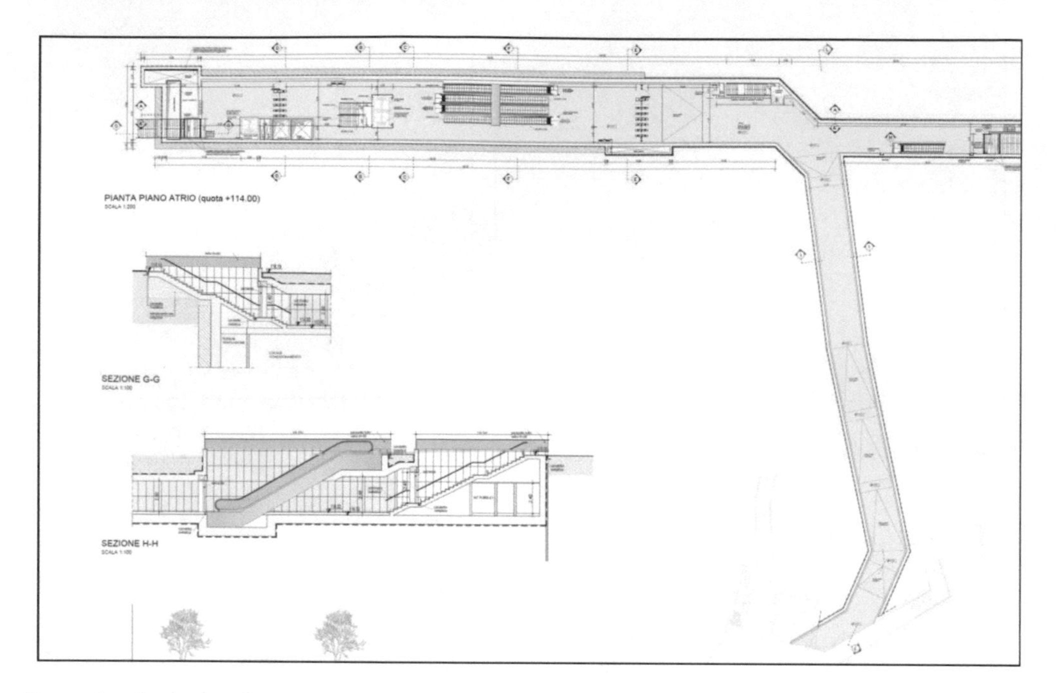

Figure 6. St. Ambrogio station – architectural plan of the new solution.

It was quite clear that the second solution would have allowed to satisfy more the requests of the various subjects involved, making full use of the design optimizations of the station body, allowing the maintenance of at least one vehicular lane for each direction on Carducci street (as required from the Municipality of Milan to alleviate the road situation), and becoming, at the same time, an opportunity to reuse and enhance the Pusterla complex, one of the monuments symbol of the area.

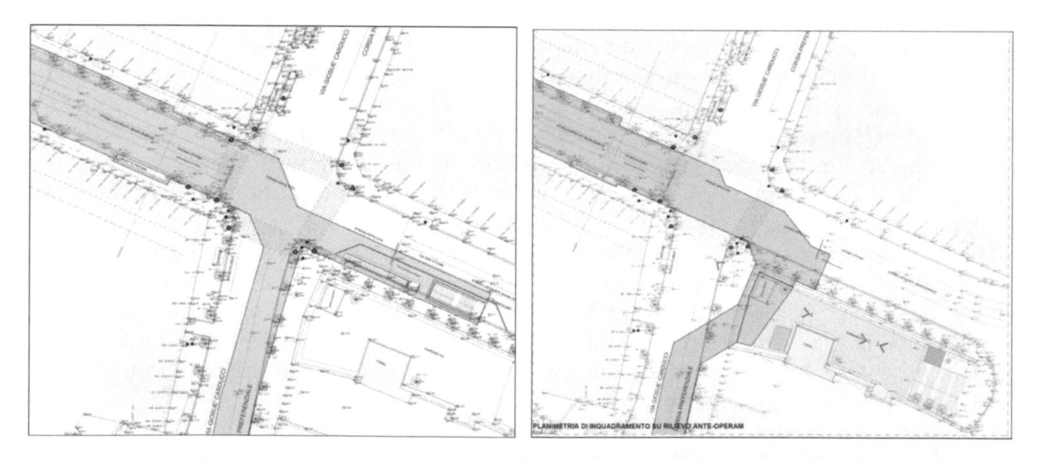

Figure 7. First (on the left side) and second (on the right side) solution of M2–M4 connection.

4.2 *Coordination with Archaeological Superintendeny and design optimization*

The desire of the Municipality of Milan to strive for the minimization of the impact of the works on the area, including within the scope of intervention an area (up to that moment in disuse) of the city with considerable historical value. such as Pusterla complex and his garden, he found the indispensable availability of the Archaeological Superintendency of Milan to evaluate its feasibility.

Figure 8. St. Ambrogio station – overview of the new solution.

Countless specific meetings were held on the topic between Municipality, Superintendency, Concessionaire, Work Supervisor and CMM4 Consortium, to define optimizations of the chosen solution. The Archaeological Superintendency, as far as it is concerned, asked for an in-depth architectural study which, starting from the analysis of the place and the value it represents, led to a project development that respected the place itself.

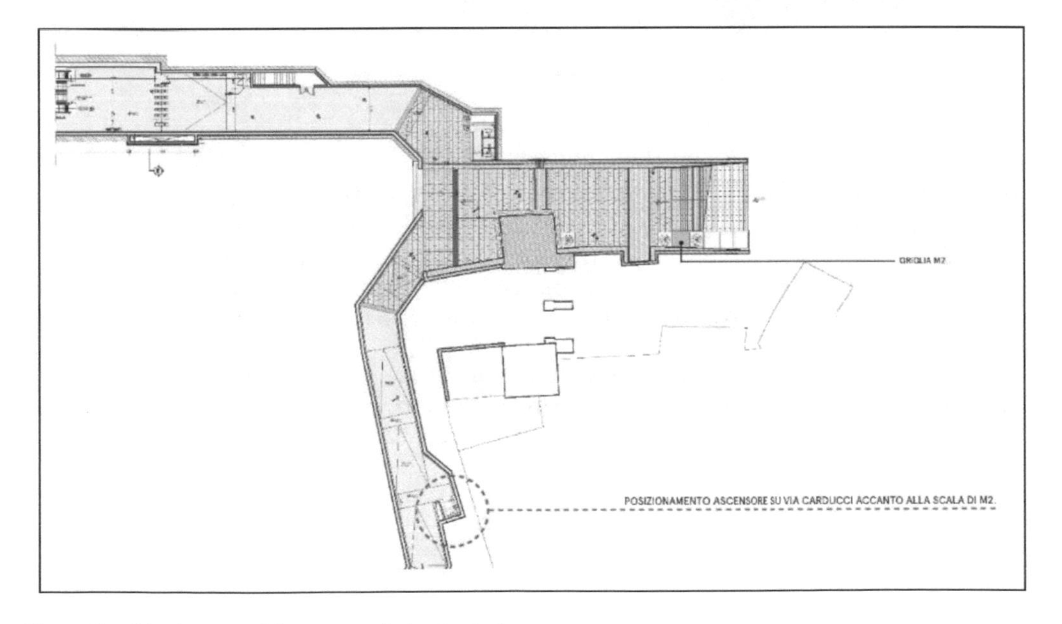

Figure 9. Planimetry of the new optimized solution.

The CMM4 Consortium, which is responsible for the Executive Design of the work (or its implementation), gave mandate, coordinating its activities, to the CREW Studio (Cremonesi Workshop), to deepen the architectural design development with the objective to combine the infrastructural needs with the stakeholders' one for a satisfactory proposal. The new concept design provides:

1. The movement of the lift at the head of the tunnel connecting M2–M4, along the Carducci street, releasing the visual cone East-West in order not to interfere visually with either the Pusterla or the entrance to the Basilica of St. Ambrogio;
2. The use of the hypogeum space along its entire length, with entrance through a monumental stairway capable of enhancing the entire architectural space and, at the same time, giving a new and less impacting seat to the pre-existing ventilation grids of the metro line M2;
3. Architectural reinterpretation of the access portal to the inter-connecting corridor (useful to reach both the M4 and the M2) of light type, with greater use of glass, material more coherent with the concepts of archaeological philology expressed by the Superintendency in cases of commingling between past and present;
4. Intervention of the perimeter walls of the hypogean garden and careful selection of the finishing materials of the area in order to highlight the historical stratifications.

Figure 10. Before work view vs night render of new optimized solution.

4.3 *Executive phases and current status of the works*

After completing the entire perimeter of the station diaphragms (in several phases) and having modified the entire network of subservices (partially in a definitive way, partially in a provisional way), after having also carried out the first digging in order to realize the consolidation of the bottom part of the station, currently the building site proceeds with the construction of the roof covering of the station, from which will continue to work in top-down mode.

The road network configuration allows, maintaining the continuity of the road system in both directions on Carducci street and the public transport, to start carrying out the preparatory activities for the execution of the tunnel connecting M2–M4.

Figure 11. St. Ambrogio station – construction phase and advancement of work.

5 CONCLUSIONS

The numerous problems linked to the Sant'Ambrogio station have made clear the need for the simultaneous involvement of all the parties involved in the construction of line 4; the mutual collaboration and cooperation allowed to re-modulate the design solution so that, in addition to solving the functional and executive needs typical of an infrastructural work, it took the opportunity to integrate and enhance a symbol of the citizen historical heritage thus contributing to the progress of the city in terms also cultural.

This fully integrated approach to design is once again a winning one and is undoubtedly a way to go with ever greater conviction and effectiveness, as well as with ever greater anticipation compared to the dynamics of project development, in search of the maximum optimization of the entire building process, with particular reference to public works, which, by their nature, anticipate a complex system of procedural management which tends to minimize, if not to exclude, the design review in the phases implementation of the work, contributing in this way to face its realization with greater awareness and serenity, respecting, also times and costs.

6 AKNOWLEDGES

All stakeholders of metro line M4 and their teams involved to find a solution on St. Ambrogio case must be thank for the hard and professional work done together, in particular Guido Mannella (president of CMM4 consortium), Massimo Lodico (CEO of Metrblu), Lamberto Cremonesi (Engineer of Crew workshop), Francesco Venza (Work Supervisor – MM spa), Antonella Rainaldi (Archaeological Superintendeny), Fabio Terragni (President of concessionaire M4) and Filippo Salucci (Municipality of Milan for M4).

Tunnels and Underground Cities: Engineering and Innovation meet Archaeology, Architecture and Art, Volume 1: Archeology, Architecture and Art in underground construction – Peila, Viggiani & Celestino (Eds)
 ISBN 978-0-367-46574-2

Tunnel warfare in World War I: The underground battlefield tunnels of Vimy Ridge, France

M. Diederichs & D.J. Hutchinson
Geological Sciences and Geological Engineering, Queens University, Kingston, Canada

ABSTRACT: Vimy Ridge was a critical battlefield on the Western Front during the first World War. The final battle (April 9-12, 1917) is a critical milestone in the building of Canada as a nation but was also a major turning point in the seemingly endless trench warfare of the day. The details of the final battle on the ground is well documented in literature, on television and in the movies of today. Less known is the underground warfare that took place at Vimy and across the front lines of WWI. Both sides used tunnelling to supply the front lines, to provide shelter for troops and to advance attacks under the enemy lines. The construction of the tunnels, the strategic design and the long- term stability and current status are matters of great interest. This work was brought about by a need to assess the potential for cave-ins or collapses during restoration works on the Vimy Ridge Monument. The tunnel layout and designs employed by both German and Allied forces are described. The tunnelling challenges of the day are discussed and the current mechanics of instability and long-term prognosis are explored along with a summary of modern site investigations and analysis. This investigation is undertaken within the important historical and human context of the tunnel network.

1 INTRODUCTION

The Canadian National Memorial, shown in Figure 1, is located at Vimy Ridge, north of the city of Arras, France (Figure 2). Occupied by German troops early in WWI, during 1916 and 1917, Vimy Ridge was tactically vital ground overlooking the critical communication centre of Arras and covering the industrial towns to the east. For most of WWI, German forces occupied and defended the Ridge and developed a formidable defensive system, protecting the northern hinge of the Hindenburg line after the German withdrawal from the Somme in early 1917.

Figure 1. Vimy Ridge Memorial site with adjacent battlefield terrain.

Figure 2. (left) Location of Vimy Ridge along the Hindenburg Line of WWI; (right) Canadian troops occupying German positions after the battle of Vimy Ridge (https://www.collectionscanada.gc.ca).

At the Battle of Arras on 9 April 1917, the Canadian Corps, comprising all four Canadian Divisions and the fifth British Division, were given the crucial task of seizing Vimy Ridge. This they achieved with tactical skill and courage after 3 days of heavy fighting against a tenacious defence, forcing the Germans to retreat. The Battle for Vimy Ridge was the first time that all four Divisions of the Canadian Corps had taken the offensive together and the achievement was subsequently hailed as a defining point in the creation of a sense of Canadian identity. The Memorial Site is situated on 91 ha of land granted by France to Canada in 1922, in recognition of the victory. The Memorial is dedicated to the 66,000 Canadian troops who died during the war, and the names of the 11,285 Canadian soldiers who died with no known grave are inscribed into the limestone around the base (http://www.veterans.gc.ca/eng/remembrance/history).

2 CONSTRUCTION OF WWI UNDERGROUND EXCAVATIONS

During WWI, the Allied and German troops faced each other from trenches located across a narrow strip of land known as No-Man's Land. A network of long subway tunnels, driven from a distance of several kilometres away constructed by hand and used to transport equipment, materials and troops to the front line (Figure 3) in the relative safety afforded by covered, underground access. Shallow communication tunnels were constructed at shallow depths under very thin crown pillars. Troops found shelter in larger dugouts excavated adjacent to the network of trenches. The type of construction depended upon the depth below

Figure 3. (left) A soldier resting during tunnel construction; (right) Supplies and troops entering a tunnel enroute to the front. (https://www.collectionscanada.gc.ca).

surface and the earth material into which the structure was excavated. The deepest of these dugouts were built to withstand direct shellfire and therefore needed an adequate cover of undisturbed rock overhead, usually between a minimum of two metres for light artillery fire and 16 metres to withstand heavy fire.

These excavations were generally accessed via steeply dipping tunnels or shafts excavated from within the network of trenches. (Figure 4). Tunnels ranged in dimensions (Figure 5). Front line tunnels were typically 1m wide by 2m tall. Main tunnels could widen to 1.5 to 2m in

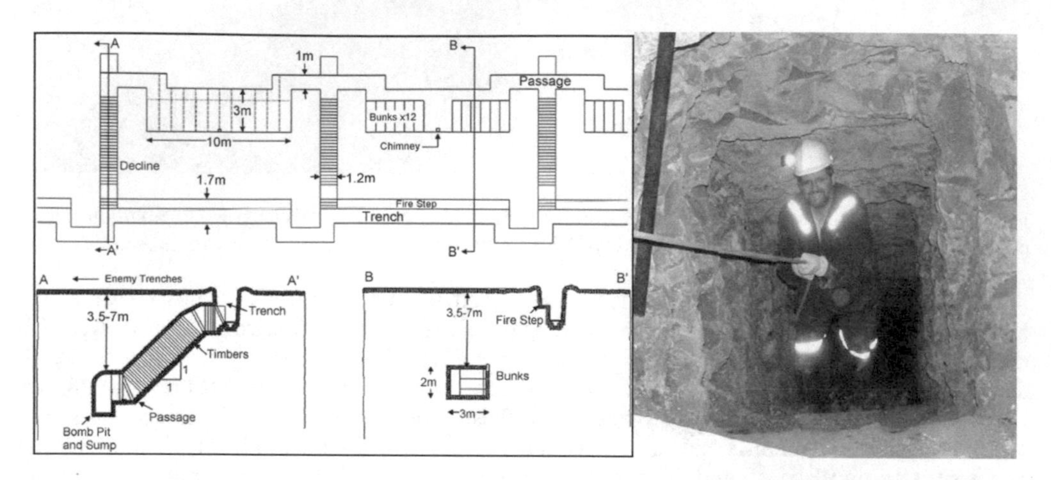

Figure 4. (left) Typical layout and section of Vimy Canadian and British tunnels (modified after Hutchinson et al 2007); (right) Typical decline into deeper tunnel system.

Figure 5. (top) Typical subway tunnels 1m wide in chalk at Vimy; (bottom) 3m wide dugout at tunnel junction.

short sections. Dugouts for sleeping and command activities were 3m wide by 2m tall. It is thought that some headquarter caverns up to 4m wide exist at depth.

On the Canadian Corps Vimy front, 13 such excavations were constructed of various lengths, totalling 10 km of tunnels as illustrated in Figure 6. The forward end of three of these (the Cavalier, the Grange and the Goodman tunnels) lie within the Memorial Site. Further description of the excavations accessed via the Grange tunnel is provided by Rosenbaum (1989).

Military mining was also used to great and terrible effect during the War. Both sides employed companies of tunnellers or sappers, often comprised of miners recruited and sent to the front lines, generally with minimal to no military training. These companies worked to excavate deeper fighting tunnels used for troop movement along and under the front line, as well as deep tunnels excavated out under No-Man's Land to locations under the enemy position. Here the sappers excavated saps, or listening posts, and underground mine chambers which were subsequently filled with explosives, to be detonated at the start of an attack on the opposing front line. The resulting massive craters as seen in Figure 5, many of which are still visible between the two front lines at the Vimy memorial site.

Many of the surviving underground tunnels, intersections and dugouts have, after more than a century of weathering and the effects of groundwater, are experiencing instability and in

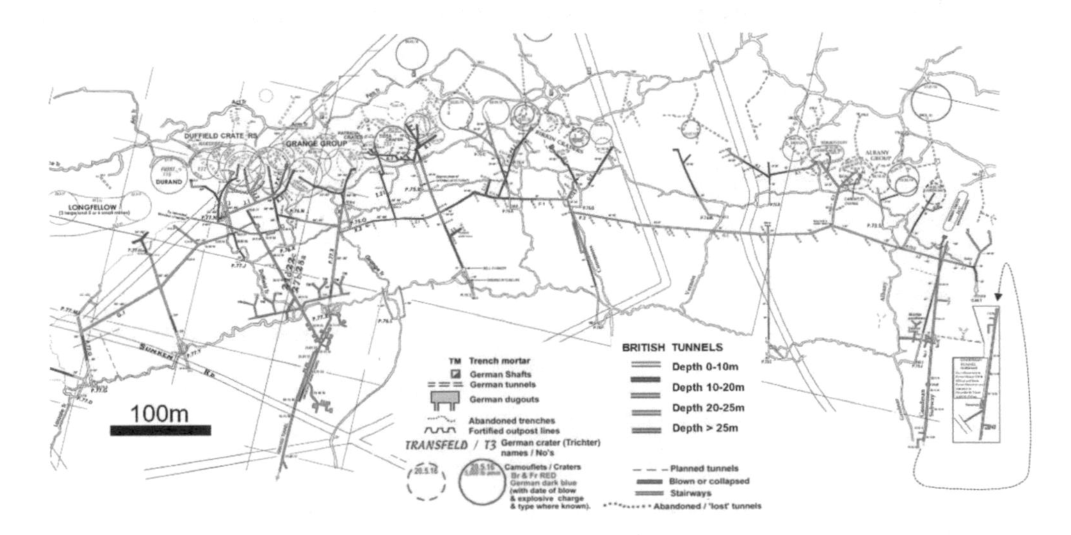

Figure 6. Part of the tunnel network ("La Folie" or P-Sector) in the vicinity of the main Grange and Goodman Tunnels. Major known explosion craters from military mining are shown as are known German tunnels. Map compiled by the Durand Group based on military maps.

Figure 7. Rehabilitation works (2006) and a typical subsidence event (right) near monument.

severe cases, caving to surface and subsidence (Dillan 2004). This subsidence poses a hazard to tourists on the site but more importantly created a challenge for the rehabilitation of the monument and the heavy equipment required for this effort (Figure 7).This paper examines this issue.

3 GEOLOGICAL CONDITIONS

The geological setting of the Memorial Sites comprises varying thicknesses of silt and clay, overlying weathered and then medium to high-density chalk. The water table is generally wellbelow the tunnels although local perching of water during heavy rains has been observed, with impact on the tunnel system.

The surficial geology of the area surrounding the Vimy and Arras area largely comprises Pleistocene silt, underlain by Coniacian and Turonian chalks. The Marqueffles Fault trends NW–SE and lies at the boundary between the syncline in the Douai plain to the north and the chalk anticline of the Artois and Somme to the south. Normal slip on this fault is responsibly for the formation of the ridge. The chalk anticline of the Artois and Somme regions is a fairly simple structure, with shallow flexural structures trending NW–SE, parallel to the Fault (Doyle and Bennet 1997). The chalk is deeply fissured, but there are no known large natural cave systems developed within the area [10]. A geological section through Vimy Ridge is shown in Figure 8.

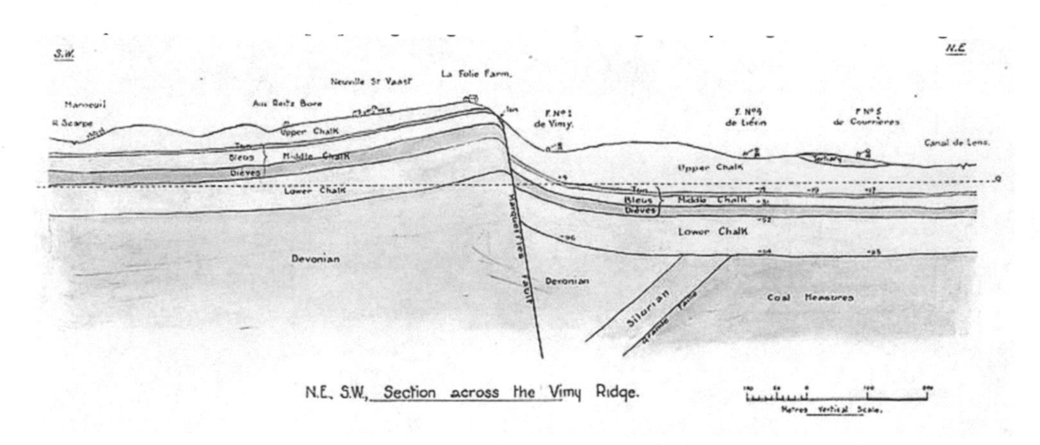

Figure 8. Geological section through Vimy ridge created in battle by Sir T.W. Edgeworth David (after Edgeworth-Davis 1917) of the Australian Tunnellers Corps.

The Marqueffles Fault has created approximately 60 m of vertical offset between comparable chalk units on either side of the fault, thereby creating the Vimy Ridge [6]. The land dips gently away to the south from the crest of the ridge at approximately 51, and is generally horizontal to the north side of the ridge in the Douai plain. The chalk is approximately 200 m thick in this part of France (Doyle 1998). Jointing in the joints were found to be orthogonal, forming cubes with horizontal and vertical faces with 0.5-1m at depth, decreasing to 10cm near surface.

4 GEOMECHANICAL PROPERTIES

Mapping and observations of the chalk characteristics and condition were made in several WWI underground excavations distributed across a large area as well as in local quarry faces (Figure 9) and in the tunnels (Figure 10). Furthermore, disturbed samples were recovered from fourteen auger drill holes (Dillan 2004) advanced to assess whether geophysical targets corresponded to subsurface voids. The ground surface on the Memorial sites has in most places been disturbed by a number of events, including shelling, excavation, subsidence, and backfilling of voids and shell holes. Furthermore, frost action has shattered the chalk such that the material in the upper few

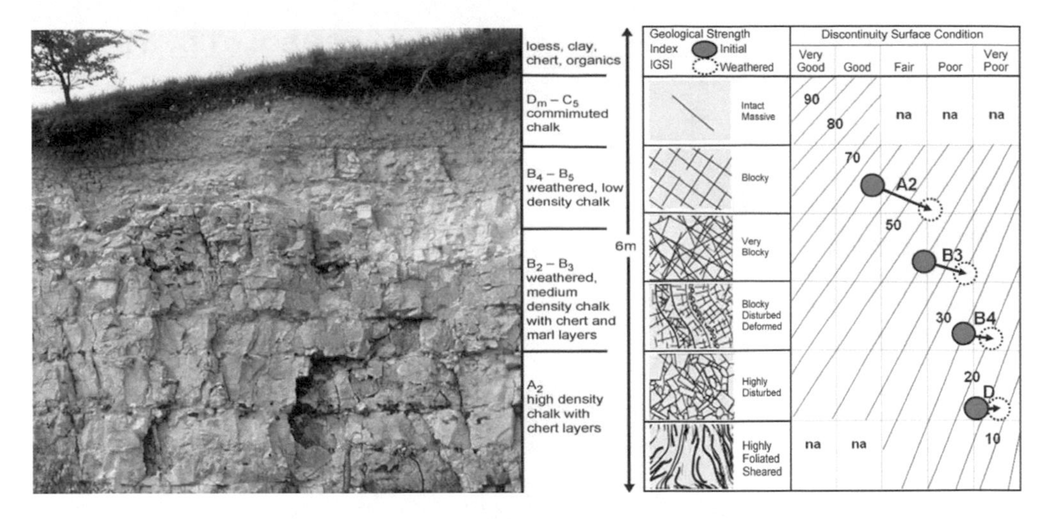

Figure 9. Chalk exposed in a local near Arras. The chalk classes are also found at depth in tunnels with less weathering based on Bowden et al (2002). The right figure illustrates estimated GSI for the chalk classes with and without weathering (modified after Hutchinson et al 2007).

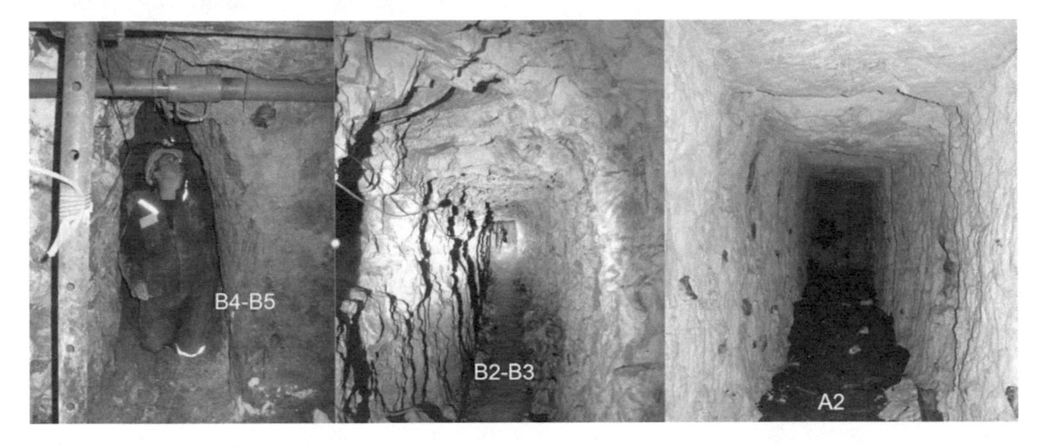

Figure 10. Underground conditions comparable to the chalk classes in Figure 9.

metres of the ground comprises floating chert nodules and chalk clasts, surrounded by silt, clay and remoulded chalk. This overburden layer is typically between 1 and 3m thick, and was found to be on average 2.6m thick during the drilling program. Weathering of the chalk has lead to dissolution features that were observed in the quarries and tunnels. The fresh intact chalk had strengths of UCS = 5 to 13 MPa and Young's modulus between 10 and 20GPa (Hutchinson et al 2007). Testing of remoulded chalk from a variety of locations, with differing calcium carbonate content and plasticity, resulted in friction angle = 31 to 33° and cohesion = 10 kPa. Typical failure modes observed in the tunnels and intersections are shown in Figure 11.

5 STABILITY ANALYSES

The delayed failures observed on the site now, approximately 90 years after the excavations were constructed, are most likely the result of progressive weathering and deterioration of the rock-mass, through progressive unravelling. The failure may form an arched or bell-shaped roof, which may become self-stable. If the arch does not stabilize before the void extends up into the disturbed and weathered chalk, the void will migrate through to surface, eventually creating a

Figure 11. Typical Failure modes observed in the Vimy Ridge tunnels (after Hutchinson et al 2007).

hole in the ground. Due to the bulking of the rockmass during failure, estimated to be about 10 to 15% (Lord et al 2002), the failure can progress without choking off to a height of 6 to 12 times the height of the excavation. Similar unravelling failure of the chalk may progress along continuous, vertical joints widened by dissolution. The potential for collapse and ground surface disturbance depends upon the span of the excavation, as well as the thickness and strength of the clay and chalk above the excavation roof. Subsidence could also result from sudden collapse of a previously effective support system (timber or crude concrete or steel beams).

Assessment of the likelihood of failure is made more difficult at this site by the long history of disturbance, including: i) from the surface, the effects of shelling, excavation, backfilling, landscaping, and focused water flow, and ii) from underground, stowing of waste or fill into underground excavations, destruction upon retreat, and the influence of previous subsidence events. However, the influence of each of these factors on the stability of the excavations, and therefore upon the possibility for subsidence failure, was explored using numerical modelling tools. The stability analyses were performed to better understand the relationships between tunnel depth, span, geological material and overall stability. The analysis was performed using 2D finite element code utilizing plasticity solutions described by Owen et al (1980) and Chen (1982). Rockmass properties were assessed according to Hoek et al (1999, 2002, 2006)

The results presented in Figure 12a are for a standard 3m wide dugout (2m high) located at 6m depth (to the roof) or, more importantly, 1m below the critical transition to fair quality blocky chalk from poor quality weathered chalk above. Figure 12b shows the results for Phase2 models of a 3m wide dugout at 3.5m, 4.5m, 6m and at 7m depth, both at the time of excavation (left) and after groundwater weathering, blast disturbance and gravity fallout of yielded material has occurred (right). Figure 13 shows underground examples of this failure mode.

Shallower excavations (<4m) present a lower hazard today, as the failure has likely already occurred, either during construction (excavation abandoned due to mining difficulties), during early service due to blast vibration, or shortly after the war. These depressions may have been filled from surface or may be visible as a depression in the ground surface. Deeper tunnels below 6.5m are less likely to fail, at any time, provided that they are more than 1.5m into the fair quality (A2) chalk.

Subways (typically 1m wide by 2m high) and declines, represent a lower hazard. This is because deeper tunnels are inherently stable, and shallower tunnels (of smaller dimension than the dugouts) will induce only a limited extent of failure due to span constraints. Total failure is likely only for subways and decline accesses that are present in the uppermost chalk layers.

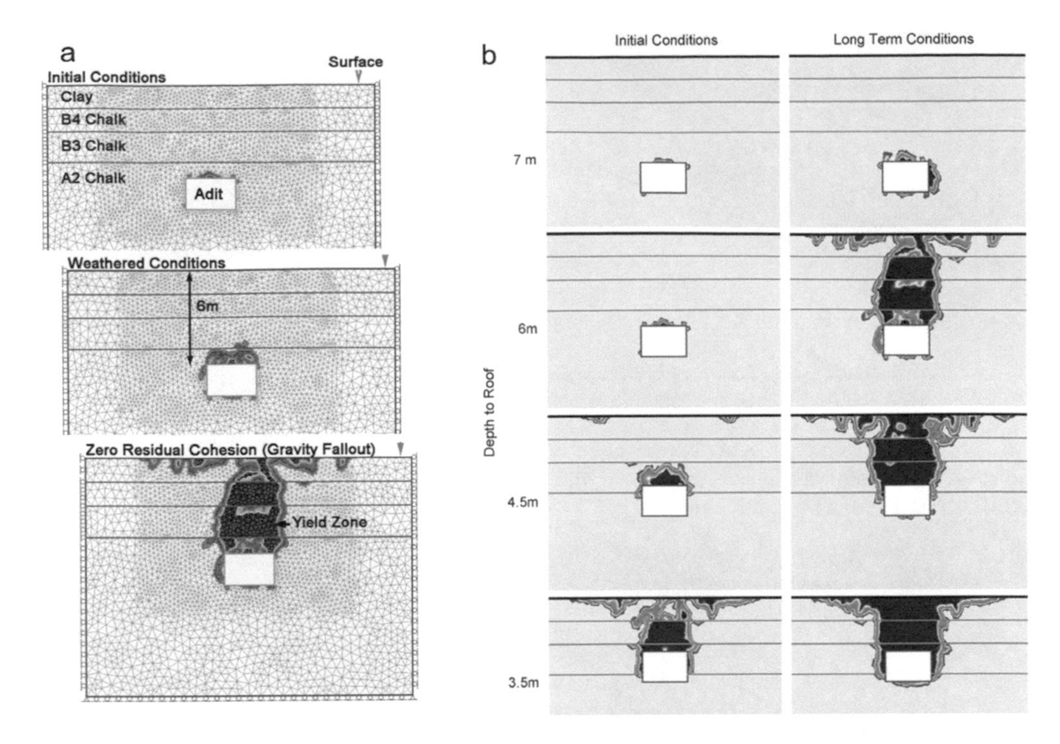

Figure 12. Examples of finite element analysis described in detail in Hutchinson et al (2007).

Figure 13. Example of progressive collapse and bell chamber formation in the Vimy tunnels.

Further analysis was conducted using the iterative Voussoir solution for a vertically jointed beam presented in Diederichs and Kaiser (1999a) and illustrated on Figure 14a for this case. The assumption is that if each individual beam is stable then failure cannot occur. If the lowermost beam fails as in Figure 15, however, then perhaps the next one will proceed to failure as well, and so on. This leads, in the extreme, to ravelling failure of the whole rock-mass above the excavation.

In order to understand the effect of dissolution by ground water percolation within the joints, a beam relaxation approach Diederichs and Kaiser (1999b) was used (Figure 14b) . In this analysis, the undeflected beam is effectively shortened by the cumulative width of material assumed to be removed from the joint surfaces by weathering. While this analysis is illustrative only, it demonstrates that for stable openings in fair to good ground, as little as 1 to

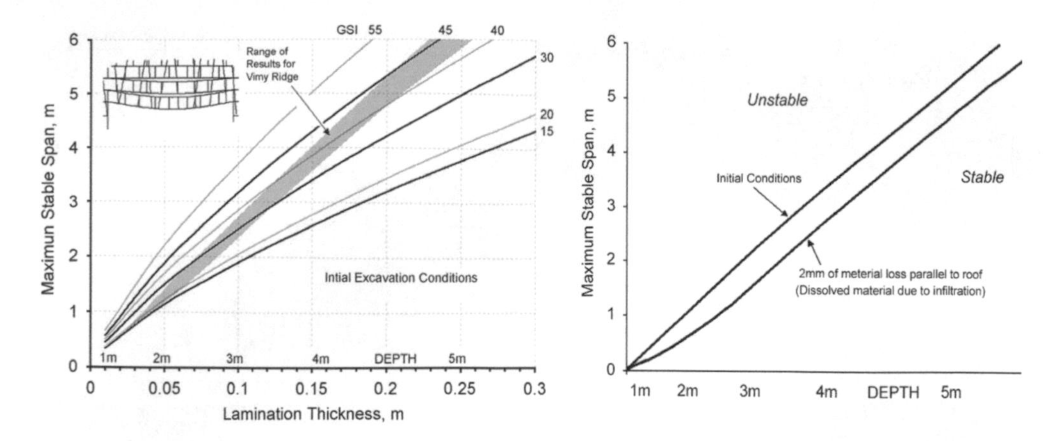

Figure 14. Voussoir beam analysis of vertically jointed, horizontally laminated chalk roofs including (right) material loss in vertical jointing.

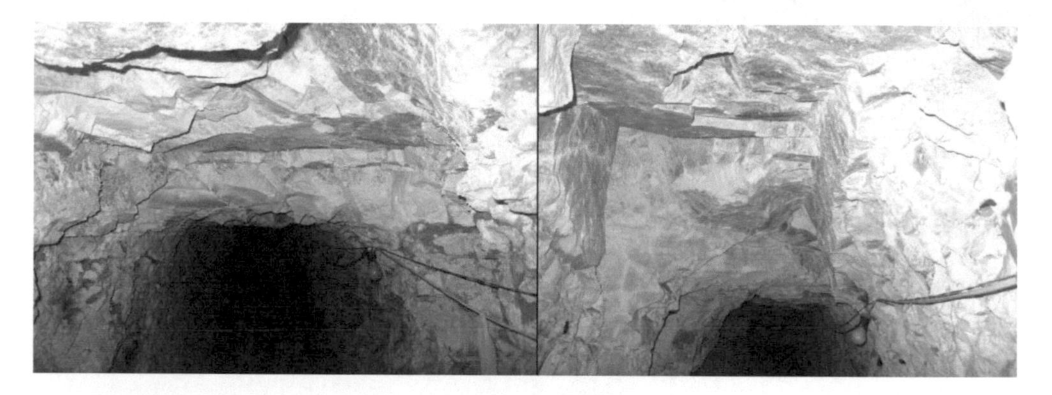

Figure 15. Failure or delaminated beams in the Vimy tunnels.

2mm of material loss perpendicular to the roof can cause failure for dugout spans in fair chalk or subway tunnels in poor chalk conditions.

6 CONCLUSIONS

1) An assessment of material properties, based on published literature and a field assessment of the material condition and properties lead to the development of an idealized geological section through the upper layers of the chalk and overburden.

2) Numerical simulation of the excavations using finite element analysis of and equivalent continuum rockmass to assess the effect of excavation depth and dimensions on stability at the time of construction, after a period of rockmass strength degradation and in the event of gravity driven fallout after support collapse.

3) Determination of the general hazard relationships (Table 1) between span and depth to isolate excavation classes that pose a long-term hazard and to reduce the priority for those that are unlikely to fail to surface and those that are likely to have failed soon after construction (to be subsequently filled with debris and collapse material).

Work is ongoing to identify tunnels from surface studies including geophyics and additional archive investigation. The rehabilitation of the monument was completed and the 100th anniversary of the battle was celebrated on site in 2016. Small groups continue to tour the tunnels and more archaeological work is being done (in addition to the Grange tunnel, already open to the public.

Table 1. A summary of the results of the numerical analyses

Probability of Delayed Caving	Subway Tunnels 1m wide x 2m high	Dugouts 3m wide x 2m high
	0-2.5m depth:Tunnels have already failed and are filled.	3-4m depth: Dugouts have probably failed. Excavation of dugouts at <3m depth unlikely.
LOW	3.5-4.5m depth: remote chance of settlement. > 4.5m depth very little chance of failure.	Depths > 8m: Major failure unlikely.
MEDIUM	2.5-3.5m depth: Tunnels 3.5-4.5m depth: Intersection spans or tunnels if geo-anomalies present.	6.5- 8m: Delayed failure possible in extreme conditions of weathering
HIGH	not applicable	4 to 6.5 m: Chance of delayed failure is very high

ACKNOWLEDGEMENTS

Many thanks go to the Durand Group and Veterans Affairs Canada who continue to explore and rehabilitate the tunnels. Thanks also to the field staff of Peeter Pehme and the engineers of Dillon. Acknowledgements go to Phillip Robinson, Maureen White and Katherine Reid for their work on this field program. We express our eternal gratitude to the Canadians who achieved this victory.

REFERENCES

Bowden, A.J., Spink, T.W., and Mortimore, R.N. 2002. The engineering description of chalk: its strength, hardness and density. *Quarterly Journal of Engineering Geology and Hydrogeology*. 35: 355–361.

Chen, W.F. 1982. *Plasticity in Reinforced Concrete*. New York: McGraw Hill. 465p.

Diederichs, M.S., Kaiser, P.K. 1999a. Stability of large excavations in laminated hard rockmasses: the Voussoir analogue revisited. *International Journal of Rock Mechanics & Mining Science* 36: 97–117.

Diederichs, M.S., Kaiser, P.K. 1999b. Tensile strength and abutment relaxation as failure control mechanisms in underground excavations. International Journal of Rock Mechanics & Mining Science 36: 69-96.

Dillon Consulting Limited. 2004. Investigation and problem definition of subterranean features: Vimy and Beaumont-Hamel Memorials – France. *Report to Veterans' Affairs Canada*, PWGSC #EN388-020003: 101 pages.

Doyle, P. 1998. *Geology of the Western Front, 1914-1918*. London: Geol. Ass. Guide #61: 79 pages.

Doyle, P., and Bennett, M.R. 1997. Military Geography: Terrain Evaluation and the British Western Front 1914 – 1918. *The Geographical Journal*, 63 (1): 1-24.

Edgeworth David, T. W. 1917. N.E., S.W., section across the Vimy Ridge: *MAP Edgeworth David Coll/ 10*. http://nla.gov.au/nla.map-edlc1-22

Hoek E, Carranza-Torres CT, Corkum B. 2002. Hoek-Brown failure criterion-2002 edition. In: *Proceedings of the Fifth North American rock mechanics symposium*, Toronto, Canada, Volume 1: 267–73.

Hoek, E., and Brown, E.T. 1997. Practical estimates of rock mass strength. *International Journal of Rock Mechanics and Mining Science and Geomechanics Abstracts*, 34 (8): 1165-1186.

Hoek, E., Diederichs M.S. 2006. Empirical estimation of rockmass modulus. Int J Rock Mech Min Sci. 43: 203–215

Hutchinson, DJ., Diederichs, MS., Pehme, P., Sawyer, P., Robinson, P., Puxley, A. and Robichaud, H. 2007. Geomechanics stability assessment of WWI military excavations at the Canadian National Vimy Site. *International Journal of Rock Mechanics & Mining Science*. v45:1:p59-77.

Lord, J.A., Clayton, C.R.I., and Mortimore, R.N. 2002. *Engineering in Chalk. London*: CIRIA.

Masefield, J. 1917. The Old Front Line. London: William Heinemann. 128 pages.

Owen, D.R.J., and Hinton, E. 1980. *Finite Elements in Plasticity*. Swansea: Pineridge Press: 589p.

Rosenbaum, M.S. 1989. Geological influence on tunnelling under the Western Front at Vimy Ridge. *Proc. Geol. Ass.*, 100 (1): 135-140.

Tunnels and Underground Cities: Engineering and Innovation meet Archaeology, Architecture and Art, Volume 1: Archeology, Architecture and Art in underground construction – Peila, Viggiani & Celestino (Eds)
 ISBN 978-0-367-46574-2

Archaeology in underground construction: The experience acquired during construction of Italian high-speed railway lines

F. Frandi
Italferr Spa, Roma, Italy

ABSTRACT: Since 1996 FSI (Ferrovie dello Stato Italiane – Italian State Railway) was the first client in Italy to set up a team of archaeologists, for the purpose of encouraging a new approach to the archaeological theme. Archeology becomes an essential step in the design process of the strategic infrastructures, bringing new awareness and sensitivity about the sustainability of transport networks, which transform the surrounding environment. This new approach has a double purpose: to increase the Italian transport infrastructure network and to preserve and enhance the archaeological heritage. During the works of the underground construction for the High-Speed line, a significant part of the eastern suburbia of the ancient Rome was discovered. A suburban necropolis – one of the largest found in the last decades – has been excavated together with a part of the *via Collatina*. Once again along, the investigations unearthed one of the largest *fullonicae*, a unique complex for extension, plan and conditions.

1 ARCHAEOLOGY IN THE CONSTRUCTION OF HIGH-SPEED RAILWAY LINES

The FSI Group experience in the archaeology field, on occasion of the construction of the High-Speed lines, and the legislative proposal deriving therefrom inspired the law-maker to institutionalize the Archaeological Preventive Evaluation, subsequently transposed by Articles 95 and 96 of the Public Procurement Code (Legislative Decree No. 163/2006) and by its further revised version under Article 25 of the Legislative Decree No. 50/2016.

The Preventive Archaeology under the Public Procurement Code and further enclosed with the project at its various development stages prevents the extension of the completion time needed for the works following the archaeological discoveries and the significant increase of the related costs.

FSI Group was the first Client in Italy that in 1996 equipped itself with a team of archaeologists intended to put in place a new approach for the Preventive Archaeology matters.

Due to their expertise acquired through training, Italferr, the Engineering Company of FSI, with dedicate working unit, carries out the activities related to the Archaeological Preventive Evaluation such as studies, planning and execution of archaeological investigations, enhancing of the archaeological findings, by paying attention to:

- the optimization of the interaction between the archaeological heritage and the infrastructure development;
- improve the quality of the projects;
- improve the sustainability of transport networks;
- a better management of impact resolution;
- the time-cost streamlining in realizing the infrastructures.

More specifically, Italferr, in collaboration with the competent local and central heritage authorities, carries out throughout Italy:

- the drafting and the verification of archaeological studies;

- the planning of the archaeological investigations;
- execution of archaeological drilling and archaeological excavations, as a part of the different investigation step of the Archaeological Preventive Evaluation;
- enhancing of the archaeological heritage in areas crossed by the infrastructures.

The outcomes of the archaeological studies and field investigations become immediately an integral part of the railway projects, since the beginning of their drafting, with the aim to resolve the impact between the ancient evidences and the new infrastructures and the civil works, granting the permissions required to the start-up of the construction stage.

This new approach allows Italferr to reach significant goals, such as the almost complete exclusion of the archaeological discoveries in progress, the reduction of the costs by 50% and the decrease by 40% in the time needed for the archaeological investigations, as compared to the High-Speed line experience.

About the Preventive Archaeology, Italferr is the most structured and organized Company amongst the large Contracting Authorities that operate at the national level.

During the development processes, archaeology becomes a part of the project that gives new awareness and sensitivity to the strategic infrastructures intended to bring significant changes to the territory. The Company has a twofold objective: the first one is to complete fundamental works for the development of the Country (see, the High-Speed Lines that changed the Italians' travelling customs with undoubtful advantages for the whole community), and the second one is to protect and to enhance the huge archaeological heritage that encloses the historical identity and heritage of the whole Peninsula, also for the future generations.

The archaeological activities carried out by Italferr, completely enclosed with the planning stages, prove the company's expertise to put in place the necessary multi-disciplinary approach. The team of archaeologists actively collaborate with the other experts involved in the planning and the execution of the activities.

Due to the synergy created among the different expertise, Italferr draws up innovative design solutions intended to provide the useful items for the development of the most appropriate design solutions, with the aim to protect and at preserve any ancient environments found during the archaeological investigations.

Within the post-excavation activities, Italferr promotes the enhancing activities for disclosing the discoveries, with the awareness that the construction of the new infrastructures represents a unique opportunity to improve understanding and fruition of our cultural heritage.

Currently, Italferr is deeply engaged in the activities derived from the "Sblocca Italia" Decree (Law No. 164 of 11 November 2014), which sets forth the time streamlining for the approval and implementation of the Naples - Bari and Palermo – Catania Railway Lines. The recent archaeological activities recorded a significant economic increase with an expenditures incurred in the last three-years of around 20 million Euros.

During the investigations along the Naples - Bari railway, 373 areas were subject to archaeological digging over an overall area of 580,000 sqm and 350 workers were employed as labor force.

Currently, on the Naples - Bari line, ancient environments such as villages and pre- and proto-historical necropolis, settlements and funerary areas of the Italic (*Dauni, Sanniti* and *Campani*) populations, environments dating from the Roman age, such as villas, roads (*via Appia Antica*), necropolis and hydraulic systems are being investigated and enhanced. The valorization of the most significant discoveries requires restoration works, expert analyses, anthropological studies, 3D reconstructions, museum-related operations (among the most recent ones, the painted grave belonging to the Campanian culture).

2 ARCHAEOLOGY IN THE UNDERGROUND CONSTRUCTION

During the underground works for High Speed in Rome, the significant discovery of the Western part of the ancient city is particularly important.

A suburban necropolis – one of the largest found in the last decades – has been excavated together (Buccellato et Catalano 2003; Buccellato 2006; Buccellato et al. 2008; Musco et al. 2001; Musco 2001) with a part of the *via Collatina*, an ancient roman paved road (Figures 1, 2,3,4).

This discovery allowed a new archaeological and topographical reading of the landscape, especially concerning the road track which ran from the Palatine Hill to the ancient latin city of *Collatia*, during the roman monarchy.

The ancient *via Collatina* crossed a valley floor from East to West, being bordered the *Aqua Virgo* aqueduct to the South. Its width was limited, with a carriageway of only 2.40 m entirely paved with blocks of leucitic basalt.

Both sides of the road were provided with *crepidines*, the sidewalks, which were separated by big travertine elements with rounded tops, probably to prevent the carriages from going on the sidewalks; these were real bollard.

The diggings brought to light a large necropolis, previously unknown, aligned along the road line and characterized by a thicker concentration of burials at the intersection of *Collatina* with two other road trails. The latter two were perpendicular to the ancient via

Figure 1. *Collatina* and necropolis (*Cf. References*).

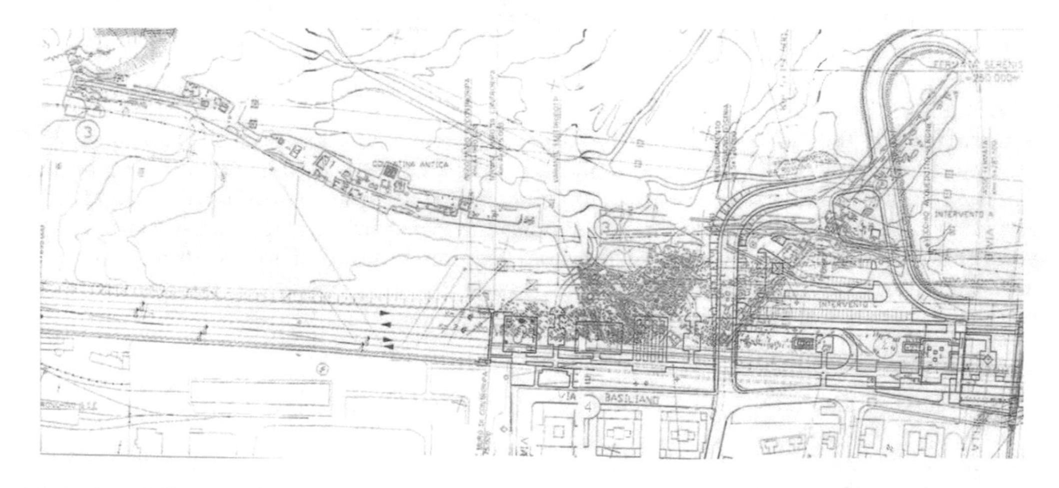

Figure 2. *Collatina* and necropolis (*Cf. References*).

Figure 3. *Collatina* (*Cf. References*).

Figure 4. *Mausoleum* (*Cf. References*).

Collatina and intended as longitudinal connections between *via Tiburtina* in the North and *via Prenestina* in the South, used until the Republican age.

More than 2.200 tombs were excavated (Figures 5,6,7,8).

The large necropolis consisted of funerary buildings, among which some monumental mausoleums, and pit graves, burials hollowed out of a bed of rock, in addition to a few of incineration graves.

Figure 5. Funeray ara (*Cf. References*).

Figure 6. Cinerary urn (*Cf. References*).

Figure 7. Slab of a tomb (*Cf. References*).

Figure 8. Jewels (*Cf. References*).

The funerary buildings were of different types. They consisted of small units built in *opera reticolata* of 3x4 m enclosing only one burial place, rather similar to funerary enclosures. Then, there were units a little bit larger that enclosed graves and real mausoleums that appeared as large monuments enclosing a large number of graves.

Instead, the underground graves were marked on the surface by *tubuli*, marble panels, small engraved marble or travertine memorials and by monumental graves, such as the so-called a *cupa* or a *bauletto* graves, which were small rectangular masonry frames covered by red painted *cocciopesto*.

The most frequent type of graves are the pit graves, covered by tiles and bricks positioned horizontally or *alla cappuccina* and, sometimes, elevated by a mound of tuff pieces.

This necropolis was used between the second half of the first century A.D. and the end of the following century.

The *Collatina* necropolis is also a very important source of information about the funerary rituals and the life and health conditions in Rome during the imperial age.

The necropolis shows a significant use of inhumation (92%) in relation to the incineration (8%) and of the single burials as compared to the twin, multiple or collective burials (Buccellato et Catalano 2003; Buccellato et al. 2008).

No particular discovery as to the laying and direction of body depending on gender, age or death was made. Almost all individuals were laid in a supine position, with their upper limbs positioned along the body or flexed on the abdomen; the lower limbs were almost always outstretched and close to each other.

As to the age at death, it was discovered that around 14% of the population died as children, while only 10 % of the population exceeded 40 years of age and 30% died between their childhood and 20 years of age. The highest mortality rate was between 20 and 40 years of age.

Additionally, the findings seem to suggest worse life conditions and shorter lives for women. The average height was 167 cm for men and 156 cm for women. About the illnesses: 70% of the dead suffered from dental cavities and 30% from abscesses. One woman was found to have had a golden dental restoration.

Not far away, at the other gallery, the preventive investigations found a unique complex by its extension, structure and conditions (Musco 2006; Musco et al. 2008). The archaeological discoveries suggest it might have been a dye-workshop/tannery dating from the second century A.D., one of the largest known to this day. The *fullonica*, extended for about 1000 m^2, consisted of 97 basins and three huge bathtubs made in *cocciopesto* for the processing of leathers and fabrics (Figures 9,10).

Inside a large room there were a series of small square cells with a 1.5 m side, a floor covered with a series of circular terracotta tubs inserted within, operational to the processing of the fabrics and of the leathers. Some basins were designated for the pressing, while others for the preservation of the substances needed for the processing.

The basins' room overlooked a portico supported by cornerstones on a duct provided with benches, beyond which there were 3 large quadrangular bathtubs covered in *cocciopesto* and connected by terracotta tubules, which, in turn, were provided with elevated benches.

Another duct ensured the water discharge and was closed by a wooden barrier. This duct flanked another two rooms provided with a series of masonry cells, where there were tubs and presses separated by small walls.

The water supply of the building had to be ensured both by the flow of the local waters and by the secondary branch of the *Aqua Virgo* aqueduct.

The system was also provided with a *doliarium*, extended over a surface of around 450 m^2, where 44 *dolia* were discovered, fragments for the most part (Figures 11,12,13).

The system discovered could have been used as a *fullonica*, namely a dyeworks/tannery, or as a *officina coriariorum*, namely a tannery. The arrangement of the rooms, in fact, fit both factory types. In particular, the first basins' room, where the tubes are fenced by very small walls, seems more like a tannery. The three bathtubs and the other rooms with *dolia*, fenced by higher walls, seem more like a *fullonica*. This hypothesis may confirm the discovery of a rope and of some linen fabric fragments inside the concrete mix.

The system was positioned towards the Easter-Western side and was adjacent to the ancient *via Collatina* route, which was discovered during the diggings, overlooked by 7 funerary buildings.

Next to the *fullonica* there emerged the route of the ancient *via Collatina*, about 4 m wide as well as 5 *columbaria*, with graves dating back to the late roman republican age (Figure 14).

Figure 9. *Fullonica*/tannery, via *Collatina* and necropolis (*Cf. References*).

Figure 10. *Fullonica*/tannery, via *Collatina* and necropolis (*Cf. References*).

Figure 11. *Fullonica*/tannery (*Cf. References*).

Figure 12. *Fullonica*/tannery and necropolis (*Cf. References*).

Figure 13. *Fullonica*/tannery (*Cf. References*).

Figure 14. *Via Collatina and* necropolis (*Cf. References*).

3 CONCLUSIONS

FSI was the first Group in Italy to have a team of archaeologists, to enhance Archeology as an essential step in the design process.

The combination of infrastructure and archeology acquires a new and positive value, aimed at turning archeological findings into design parameter, spreading culture and ethical and civil sensitivity.

REFERENCES

Buccellato, A. & Catalano, P. 2003. Il comprensorio della necropoli di via Basiliano. In Mélanges de l'école française de Rome: 311–376. Roma: École Francaise de Rome

Buccellato, A. 2006. Municipio VI: riti e contesti funerari. In Tomei, M.A. (ed.). Roma. Memorie dal sottosuolo. Ritrovamenti archeologici 1980/2006: 329–342. Verona: Electa

Buccellato, A. & Catalano, P. & Musco, S. 2008. Alcuni aspetti rituali evidenziati nel corso dello scavo della Necropoli Collatina (Roma). In J. Scheid (ed.). Pour une archéologie du rite: nouvelles perspectives de l'archéologie funéraire: 59–88. Roma: École Francaise de Rome

Musco, S.& Petrassi, L. & Pracchia S. 2001. Luoghi e paesaggi archeologici del suburbio orientale di Roma. Roma: Ministero per i Beni e le Attività Culturali. Soprintendenza Archeologica di Roma

Musco, S. 2006. Tra via Tiburtina e l'autostrada Roma-Napoli. L'attività della Soprintendenza Archeologica di Roma. In Tomei, M.A. (ed.). Roma. Memorie dal sottosuolo. Ritrovamenti archeologici 1980/2006: 278–328. Verona: Electa

Musco S. & P. Catalano & A. Caspio & W. Pantano, K. Killgrove 2008. Le complete archéologique de Casal Bertone. In J. Faton (ed.), Dossiers d'archeologie. Dijon: Editions Faton

Tunnels and Underground Cities: Engineering and Innovation meet Archaeology, Architecture and Art, Volume 1: Archeology, Architecture and Art in underground construction – Peila, Viggiani & Celestino (Eds)
 ISBN 978-0-367-46574-2

The architecture of underground dwellings in Iran

S. Hashemi
Iranian Tunnelling Association (IRTA), Tehran, Iran

ABSTRACT: Caves were the first dwelling of humans and his experience with architecture. However one needs to make a distinction between natural and manmade caves. The practice of creating buildings and other structures by carving natural rock is known as rock-cut or cliff architecture which differs from normal architecture in that the type and quality of material being used cannot be chosen. Hence the freedom available in normal architecture does not exist. Based on construction type, cliff architecture can be classified into several main groups. Studying these dwellings provides a historic insight and information on the evolution of culture, art and architecture of different eras. This paper aims to review the different types of underground structures in addition to the historic background and architecture of some ancient underground dwellings in Iran.

1 INTRODUCTION

Using underground spaces dates back to the early history of mankind. The first inhabitants of underground spaces were troglodytes or cave-dwellers. Caves were the first dwellings of humans and his first experience with architecture. Even though it is assumed that mankind has abandoned caves for geographic mobility and the greater expedience of surface dwellings, caves have enjoyed continual habitation since those early times. However one needs to make a distinction between natural and manmade caves. Due to their special position many natural caves were used for living and defensive purposes after some changes were carried out in them. Such modifications or artificially created caves were man's first architectural endeavors by which he was enabled to create his own dwelling with a personal and spontaneous freedom of expression. Hand-dug and Rock-cut troglodytic architecture differs from normal architecture in that the type and quality of material being used cannot be chosen. All that is available is the mound to be dug or the cliff to be carved and shaped. Hence the freedom available in normal architecture does not exist. The design of access, lighting, ventilation and the nature of the subsurface structure itself depends on site specific circumstances such as the geographic location, topography of the area, and the available material. Climate conditions determine the overall appropriateness of the underground structures but the geological conditions, amount of precipitation, and defensive requirements will dictate the relative feasibility of the development of such structure. In addition to the above, cultural aspects will also contribute to the development of such dwellings which can be seen in some decorative details or patterns of community relationships.

The above aspects can be seen in various underground dwellings existing in Iran and numerous classifications can be suggested for these dwellings based on their external appearance, their interior and their usage (Labs 1976; Ashrafi 2011; Hashemi 2013; Mohamadifar & Hemati Azandaryani 2017). Although it should be mentioned that other types of underground structures such as underground temples and tombs also exist, the main focus of this paper however is to review different architecture of sub-surface dwelling types. Based on construction type, underground dwellings can be classified into the following three main groups:

a) Dwellings in which spaces are excavated in large cliffs where each space becomes a separate unit that could include windows, louvers and ornaments. The Kandovan village in Iran is a fine example of such architecture.

b) Structures where holes and cavities inside mountains and hills were developed and expanded by man with no possibility to construct ventilation systems, windows or to include exterior decorations. Hilevar and Meymand villages are such examples.
c) Underground spaces that were entirely cut into cliffs or mountains by man in order to create suitable living places. Some underground spaces formed below hills and mounds contain an inter-connected labyrinth of corridors and wells in addition to living chambers as a sort of protection against invaders and enemies. The underground city of Noushabad is an example of such spaces.

Some of the characteristics of the above mentioned dwellings are explained in the following sections. Studying these dwellings provides a historic insight and information on the evolution of culture, art and architecture of different eras.

2 KANDOVAN VILLAGE

The Kandovan village, also referred to as Kandoujan by locals, is located about 50 km south west of Tabriz and 20 km south east of Osku at the foothills of Sahand Mountain in the province of East Azarbaijan (northwest of Iran). This village is a wonderful example of man-made cliff dwelling which is still inhabited (ICHO 1997). It extends in east-west direction along the Kandovan River for approximately 5 km. Its elevation increases from west towards east to maximum 60 m from the bottom of the foothills.

The first inhabitants of Kandovan village were probably the residents of Hilevar village which escaped to this place to protect themselves against the Mongol invasion in the 13th century. In past times the fields opposite of Kandovan were the summering place of nearby villages. Pieces of pottery show that people resided in the fields up to the Ilkhanid era. Since the migrants found this place suitable for defense against invaders, and the cone-shaped cliffs for construction of dwellings, they permanently settled in that area and houses were gradually cut into these rock cones. The troglodyte homes, excavated inside volcanic rocks and tuffs, are locally called "Karaan" (Figure 1). Approximately 100 dwelling units exist in this village.

With increasing population more and more Karaans were cut into the Lahars (volcanic mudflow or debris flow) of mount Sahand. The cone form of the houses it the result of lahar flow consisting of porous round and angular pumice together with other volcanic particles that were positioned in a grey acidic matrix. After the eruption of Sahand these materials were naturally moved and formed the rocks of Kandovan (Taghizadeh & Manafi 2008).

Figure 1. Outer view of cliff dwellings called Karaan in Kandovan.

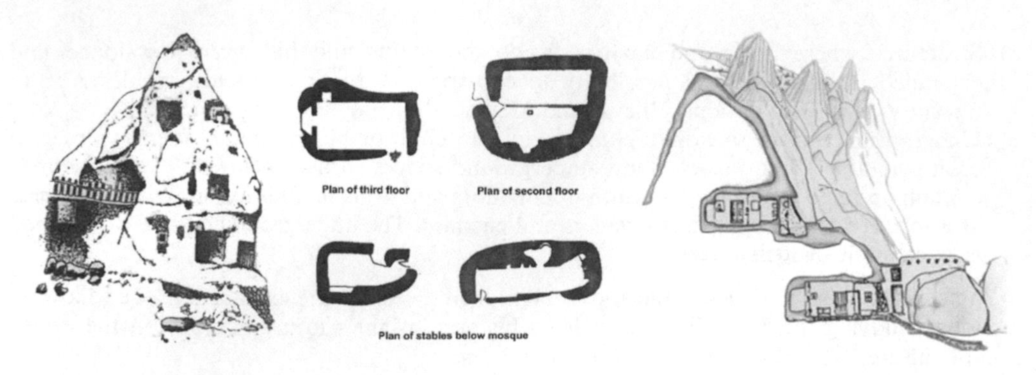

Figure 2. Cross Section and and floor plan of two Karaans in Kandovan.

Around the village the thickness of this formation exceeds 100 m and with time due to water erosion the cone shaped cliffs were formed. Various grooves that were actually natural water passages have separated the cone-cliffs from each other and formed the current streets of the village. Over some of these streets, bridges have been built that connect two Karaans to each other. The Karaans are developed in east-west direction and are rather concentrated on the eastern side, whilst most entries and living units have a southward direction. Some Karaans are as high as 30 to 40 m.

The inside space of the dwelling units is cut into the rock as much as needed and possible. The space is sometimes divided into several parts (Figure 2). Nowadays new structures made of rocks, wood, bricks and mortar are also constructed in the village and existing spaces are extended.

To form the spaces inside the cliffs the natural form and properties had to be considered. The spaces are quite simple and satisfy only the basic needs. Depending on the size and volume of the rock cones, one or more levels could be cut into each Karaan. These spaces are directly connected to the outside as there is no connection between levels from the inside. Most Karaans consist of two levels but some have three or even four different levels. The first levels have a larger area but less light and are mostly used as stables due to easier access. Upper levels are mostly used as living space. These were cut in form of square shaped rooms that sometimes include small storage chambers. In addition to entries and openings in form of windows, benches and niches were also cut out of the rock. Cutting windows out of the rock was easier in upper levels due to smaller diameter of the cone and lesser rock thickness (Gorji Malhabani & Sanaee 2011). The type of rock and the wall thickness, help to keep the interior easily warm during winter and pleasantly cool during summer.

3 HILEVAR

This village is located in East Azarbaijan province about 2 km west of Kandovan and dates back to the pre-Islamic era (ICHO 2005).

This village is carved into one of the heights of the Sahand volcano. Due to its eruption, material such as lava, ash, and pumice covered these hills and formed the rocks of this area. The rocks are irregular and vary in hardness and strength. In some parts, cone shaped structures were formed from volcanic rock. More than 50 dwellings were constructed in the subterrain rocks, all excavated by using chip-axes. The village was named Hilevar (meaning cunning in Farsi) because the dwellings were a complete camouflage and not noticeable even from a close distance (Figure 3). The inhabitants were deserted as a result of the Mongolian attack in the 13th century A.D. and most dwellers probably moved to the nearby Kandovan plain (Homayoun 1977).

Figure 3. The entrance and interior of underground dwelling units in Hilevar.

Natural phenomena such as floods have destroyed many of the houses or covered them with mud and soil and most entries are difficult to find. In recent years some of the dwellings have been repaired and renovated. The style of architecture suggests both living and defense purposes for these dwellings. Most dwellings had a stable, as keeping their animals safe was important to the inhabitants. Some of the dwellings were built in two levels. Some of the chambers are as large as 50 m^2. The height of the chambers varies between 150 and 190 cm. Some chambers include niches for various uses. Half a meter above the floor a 30 cm groove has been carved all around the chambers on the walls. This was probably used to direct moisture and rain water to the outside. Some parts of the dwellings were damaged due to flood or other reasons. These were rebuilt by the dwellers with stone walls.

4 MEYMAND

The village of Meymand with an area of approximately 420 km^2 is located in Kerman province (southeast of Iran). It was enlisted as one of Iran's Heritage sites in 2001 (ICHO 2001) and as a UNESCO World Heritage Site in July 2015 (UNESCO 2015). The village is a self-contained area at the end of a valley at the semi-arid southern extremity of Iran's central mountains (Figure 4). A thick igneous layer that covers the village has created a natural stronghold against any infiltration into the village. The main sources of water include seasonal rivers, several Qanats, and several water springs.

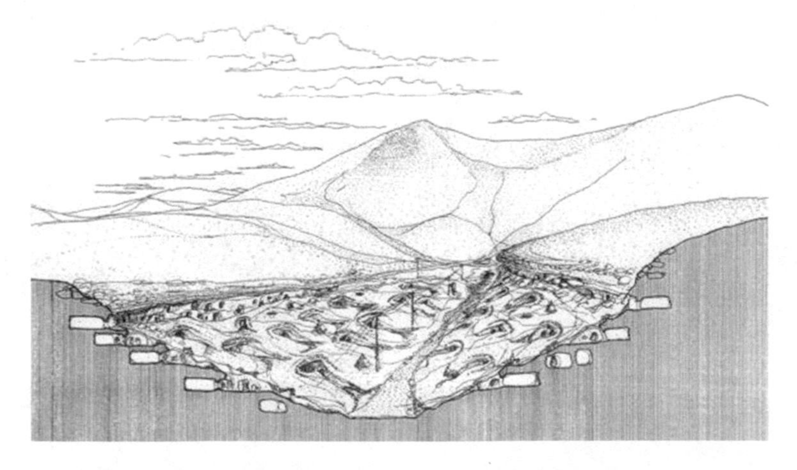

Figure 4. Cross section of Meymand with view towards the end of the village.

Meymand village is located at the intersection of two steep hills with a height of 60 m from the Talweg to the hill crest. These conditions have created a suitable place for dwelling. The living units were hewn out in form of caves inside the hills (Figure 5). It is one of the oldest dwellings in which people still continue living. The village is described as follows: "Meymand is an old famous village in Shahr-e-Babak. All rooms and chambers are excavated in the mountain foothills. All houses are constructed inside the mountain and consist of 3 to 4 rooms. Nearly 300 small alleys are dug into the rocks. It is similar to a diagonal skyscraper excavated inside the mountain, thousands of years ago" (Bastani Parizi 1945).

Two theories have been suggested regarding the origin of these structures in Meymand (Homayoun 1972). According to the first theory, this village was built by a group of the Aryan tribe about 800 to 700 years B.C. and at the same time with the Median era. It is possible that the cliff structures of Meymand were built for religious purposes. Worshippers of Mithraism believe that the sun is invincible and this guided them to consider mountains as sacred. Hence the stone cutters and architects of Meymand have set their beliefs out in the construction of their dwellings. The other theory is that the village dates back to the second or third century A.D. During the Arsacid era different tribes of southern Kerman migrated in different directions. These tribes found suitable places for living and settled in those areas by building their shelters which developed in time into the existing homes.

The Meymand village also had a defensive function. The Kerman region was attacked by different tribes over many centuries and several groups fled to Meymand which sheltered them like a fortress (Homayoun 1973).

A thick igneous layer covers the village and prevents any intrusion like a natural fortress. The builders of Meymand used the gentle slope of the mountain and primitive tools to construct a complex which finally formed a village consisting of alleys, dwellings, barns, public bath, etc. More than 400 man-made residential units are still the home of many locals. These dwellings are scattered across the hill slopes and built in two to five different levels (Figure 6).

Figure 5. Scenes of dwellings dug into rocks at Meymand Village.

Figure 6. The exterior of underground dwellings in Meymand.

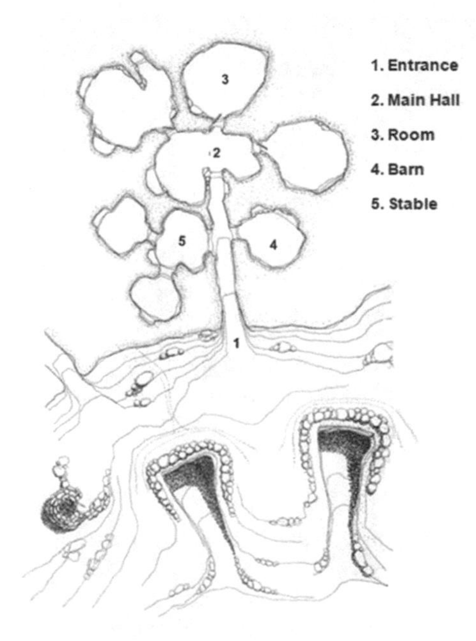

Figure 7. Layout of a dwelling unit in Meymand.

The individual cave dwelling units are about 2 meters high and have an area of 16 to 20 m^2. Small alleys known as "Kicheh" provide access to the living units, each having on average 2 to 4 rooms. Some of these units are dug up to 25 m inside the mountain. The construction of a dwelling unit starts with the chiseling of 6 to 9 meter long horizontal cuts into the cliff. The dwellings usually consist of a single square or round room (Figure 7). Most units are semi-dark as the only sources of light are the entrances. However, windows where possible were hewed openings approximately 75 cm across. Doors to the dwellings are commonly made of wood but not all of them are rectangular. Some have the shape of standing human body, narrower at the base and widening at the top to shoulder width. To prevent water ingress into the units, the threshold of the doors were raised 15 to 20 cm above the level of the Kicheh. Heat exchange in these cliff dwellings is such that the spaces are cool in summer and warm during winter.

5 NOUSHABAD UNDERGROUND CITY

The historical city of Noushabad is located northwest of Isfahan province near the cities of Aran and Bidgol and about 8 km away from Kashan. The underground part of the city served rather military and defense purposes and was excavated in three levels by hand. The first level was excavated at about 3 m below ground and the third level at a depth equal to 16 m. Each level has a height of 1.8 m. The extent of the underground city is not exactly known yet. However, archeologists guess that it has an area of approximately 15000 m^2.

Archeologists think that architectural remains of the underground part of the city maybe spread below the whole city of Noushabad and that all corridors are interconnected and lead to the outside of the city from a main passage. Their investigations also suggest that the excavation of the city started during the Sassanid period and continued during the post-Islamic era. Anthropological studies suggest that city was habited up to the Qajar dynasty (approximately 100 years ago); this was for people to protect themselves from foreign attacks and also from the heat of the desert sun. In general, influences of desert architecture such as simplicity and proportionality of the exterior with the interior, avoidance of unnecessary decorations,

Figure 8. Chambers in the underground city of Noushabad.

and abstaining from use of unnecessarily large sizes can be noticed in the architecture (Payam-e-Sakhteman 2009).

Short wells or narrow corridors would provide access to these underground spaces. The purpose of defense and refuge had largely affected the architectural plan of this city. Since access to the underground spaces should not have been easy, they were hidden in houses or fortresses outside the city borders, or inside water channels and Qanats passing below houses, gardens or Bazars, such that in case of an attack the inhabitants could quickly escape and hide from view. In some houses the access wells were hidden behind ovens (Iran Newspaper 2008).

Rooms and chambers inside the underground city were excavated with different sizes for dwelling (Figure 8). Chip-axes have left their marks on the walls. The chambers were constructed such that the connecting corridors had no direct view into the next room. About 20 cm below the roofs and with 1 m distance from each other, niches were built inside the walls for tallow-burners. This is evident from several old lamps found in these rooms which date back almost 700 years. Oil for the tallow-burners was probably produced in two ancient oil-pressing houses of Noushabad.

Based on archeological studies the different levels of this city were connected through vertical and horizontal channels. These deep shafts were also used for ventilating the space in addition to providing access paths. Some passages or channels however, did not lead anywhere and were constructed to deviate intruders or enemies and make them vulnerable. Besides all channels large stones similar to millstones were present and used to cover the entrances to the lower levels when locals were in hiding. During attacks and whilst residing in the underground spaces, water was supplied through Qanat channels.

6 CONCLUSION

Caves were the first experience of mankind with architecture. These underground spaces were used for various purposes such as living or defense. To make them more suitable for the purpose, man began to expand and change them and thereby developed the troglodytic architecture. Depending on the topography, geology and climate conditions the dwellings were constructed or shaped differently. This paper reviewed the classification of underground dwellings in terms of their architecture and several examples of dwellings in Iran were introduced. Since nowadays interest in the use of underground spaces has been revitalized due to the development of new materials and design and construction tools and techniques, studying the architectural design and construction methods of the past can provide valuable experience and help in developing new ideas for the modern world for better climate control, environmental preservation, conservation of energy and better land use planning.

REFERENCES

Ashrafi, M. 2011. "A research in Troglodytic Architecture", *Journal of Architecture and Urban Planning*, Vol. 4, Issue 7, pp. 25–48. (In Farsi)

Bastani Parizi, M.E. 1945. *The prophet of thieves.* (In Farsi)

Gorji Malhabani, Y. & Sanaee, E. 2011. "Compatible Architecture Survey with Kandovan Village Climate", *Journal of Housing and Rural Environment*, Vol. 29, Issue 129, pp. 2–19.

Hashemi, S. 2013. *The Magnificence of Civilization in Depths of Ground (A Review of Underground Structures in Iran – Past to Present)*, Shadrang Printing and Publishing Co., Tehran.

Homayoun, Gh.A. 1972. "Investigating the Meymand Village", *Historical Studies of Iran*, No. 43, pp. 119–154. (In Farsi)

Homayoun, Gh.A. 1973. "Further information about Meymand Village in Kerman" *Historical Studies of Iran*, No. 48, pp. 245–277. (In Farsi).

Homayoun, Gh.A. 1977. "The Kandovan Village", *Historical Studies of Iran*, No. 69, pp. 155–216. (In Farsi).

ICHO (Iran's cultural heritage Organization). 1997. Registration of Kandovan Village under the No. 1857 as a cultural heritage site. (In Farsi)

ICHO (Iran's cultural heritage Organization). 2001. Registration of Meymand Village under the No. 4135 as a cultural heritage site. (In Farsi)

ICHO (Iran's cultural heritage Organization). 2005. Registration of Hilevar Village under the No. 12622 as a cultural heritage site. (In Farsi)

Iran Newspaper. 2008. "The secrets of the Underground City in Kashan's Desert", No. 4011, 27.08.2008. (In Farsi).

Labs, K. 1976. "The Architectural Underground", *Underground Space*, Vol. 1, pp. 1–8, Pergamon Press.

Mohamadifar, Y. & Hemati Azandaryani, E. 2017. "A Study and Analysis of Troglodytic Architecture in Iran"; *Journal of Housing and Rural Environment*, Volume 35, Issue 156, pp. 97–110. (In Farsi)

Payam-e-Sakhteman. 2009. "Underground cities, the architectural art of Ancient Iranians", Bi-monthly technical magazine for building and construction, No 70. (In Farsi).

Taghizadeh, M. & Manafi, M. 2008. "Geotourism: The amazing Kandovan Village", GSI-Monthly, Geological Survey of Iran, No. 33, pp. 23–25. (In Farsi).

United Nations Educational, Scientific and Cultural Organization (UNESCO) World Heritage Convention (WHC). 2015. "Cultural Landscape of Maymand"; http://whc.unesco.org/en/list/1423/ [Site accessed 20/August/2018]

Tunnels and Underground Cities: Engineering and Innovation meet Archaeology, Architecture and Art, Volume 1: Archeology, Architecture and Art in underground construction – Peila, Viggiani & Celestino (Eds)
© 2020 Taylor & Francis Group, London, ISBN 978-0-367-46574-2

Spiritual life and life after death in the undergrounds of ancient Iran

S. Hashemi
Iranian Tunnelling Association (IRTA), Tehran, Iran

ABSTRACT: In some ancient cultures mountains were considered holy and the place of the Gods. Others believed that the Gods were in heaven and built their temples or tombs in higher ground to be closer to them. Irrespective of the correctness of such beliefs, the world today has inherited numerous examples of such structures and owes much of their cultural history to such cliff architecture and rock reliefs. Worshippers of ancient beliefs such as Mithraism met in underground temples which were either adapted natural caves, or artificial buildings constructed to imitate caverns. In those times, deceased bodies would be buried in a way to prevent them from coming into contact with soil, water, and fire and excavated chamber-like structures would serve as tombs. This paper aims to review the history and architecture of some of the above mentioned underground temples and tombs, constructed in ancient Iran.

1 INTRODUCTION

In the past, it was believed that mountains were one of God's first creations and in some ancient cultures mountains were considered holy and the place of the Gods. The Ark of Noah came to rest on a mountain once floodwater receded. The Ten Commandments were delivered to Moses on Mount Horeb. Abraham attempted to sacrifice his son on a mountain. Prophet Mohammed meditated on a mountain near Mecca and his first revelation was received in the cave of Hira. In ancient Greek religion and mythology it is believed that the major deities (the twelve Olympians) resided atop Mount Olympus. Others believed that the gods were in heaven and hence built their temples or tombs in higher ground to be closer to them. The most important temples constructed based on this idea are Ziggurats. These massive structures had the form of a terraced step pyramid of successively receding levels. The purpose of these structures is to get the temple closer to heavens. The Chogha Zanbil Ziggurat in Iran is one such structure that was built about 1250 B.C. during the Elamite era mainly to honor the God Inshushinak (the protector of Susa). The Egyptians built the tombs of Pharaoh's in form of pyramid buildings representing mounds which were designed to serve as stairways by which the soul of deceased pharaohs could ascend to the heavens.

Irrespective of the correctness of such beliefs, the world today has inherited several examples of structures and memorials which have withstood the hardships of nature and owes much of their cultural history to cliff architecture and rock reliefs (Hashemi 2013). In this relation, underground structures such as caves are also rich in symbolic meanings and since the past have been used for various ceremonies. The subsurface seemingly has always possessed religious and symbolic connotations. These could assist researchers, historians, anthropologists, and architects in becoming familiar with history of art and religious beliefs.

2 UNDERGROUND TEMPLES

Ancient Iranians had Mithraic beliefs even before the migration of the Aryans. In this religion Mithra, was the divinity of light and treaties and represented the sun. Followers of Mithra believed that phenomena were invincible. They believe that the sun cannot be defeated and is eternal. Mountains also were believed to be invincible and hence considered holy. Mithras is depicted as being born from a rock. It is believed that once he was born he fought and killed a sacred bull and then fired a bow at a rock to encourage water to come forth (Cumont 1903). Hence worshippers of Mithras met in underground temples called a Mithraeum. These were sometimes called Mehrab in Farsi which consist of "Mehr" meaning sun and "Ab" which means glory (and also water) (Seyed Younesi 1965). A Mithraeum was either an adapted natural cave or cavern, or an artificial building constructed to imitate a cavern to resemble the birth place of Mithras (White 1990). The cave roof resembled the vault of heaven adorned with stars. Mithraea usually face the morning sun and are commonly located close to springs or streams since fresh water appears to have been required for some Mithraic rituals.

The Aryans believed in good and evil and in their ancient religions believed in the existence of two types of gods the Ahuras (power of good) and the Divs (power of evil) which opposed each other. In the monotheistic religion of Zoroastrianism, Ahura Mazda is considered the supreme and most powerful god with other divinities beside him. Since past times fire was considered holy by ancient Iranians and is also an important element of Zoroastrian ceremonies. It was considered as an emblem of refulgence, glory and light, as the most perfect symbol of God and as the best and noblest representative of his divinity. Zoroastrians revere fire in any form and keep the sacred fire at fire-temples. As already mentioned mountains were considered a suitable place for prayers since they were close to the sky and heaven. Hence many fire temples and places of worship were constructed on top of them. To protect the sacred fire, some fire temples were built at safe and not easily accessible places such as caves.

After the advent of Islam and during Islamic rules, many of the activities of Zoroastrian and Buddhist temples stopped. Muslims at that time did not have a problem in holding their prayers in temples that was used by their ancestors for worshiping god almighty under a different name. Hence they repaired or made changes to the structures. For instance Mihrabs were added to them and in the course of time most of these structures were turned into mosques (Pirnia 1983). The following text briefly introduces two such underground places of worship.

2.1 *The Stone Mosque of Darab*

About 5 km southeast of Darab city in the province of Fars on the foothills of a low limestone ridge, an amazing structure with a cruciform design, known as Masjed-e Sangi (the Stone Mosque), can be seen (Figure 1). The structure is estimated to be more than 750 years old (ICHO 1935). This structure was also known as the fire temple of Azarakhsh. The main part of the structure is excavated inside the mountain and all its columns are made out of rock.

The Darab area was one of the religious centers of Fars province during the Sassanid era. It is argued that the stone mosque structure was initially a fire temple or the seat of a Nestorian bishop (Ball 1986, Monneret de Villard 1936, Wilber 1955). After the advent of Islam the

Figure 1. The stone mosque near Darab.

temple was turned into a mosque and an altar was carved into the rock towards Mecca (Stein 1936).

The southern side of the hill is completely dug and removed to create a flat level entrance to the Mosque. The entrance to the mosque lies on the south-eastern face of the mount. Through an arched doorway, 4.70 m high and 3.15 m wide, a vaulted narthex, a tripartite hall measuring 9.70×2.75 m, can be found. This leads into a large cruciform shaped room, measuring 16.76×15.85 m along the arms of the cross (Figures 2 and 3). The quadruple shaped plan of the interior has four tunnel-vaulted halls (transepts) around a central square. The ceiling, admitting light into the interior, consists of an entirely cut vertical shaft measuring 3.10 m^2 and is 3 m away to the upper surface of the mount. The excavation of the mosque seems to have had started top down from this shaft and then continued to the shaft walls in four directions (Bier 1986). Directly beneath this shaft opening; there exists a square shaped water pool almost the same size as the opening above. Narrow, low corridors (1 to 1.25 m wide and 1.80 m high) with flat ceiling, accessible from the lateral parts of the narthex, run behind the four halls all around the mosque, stopping short only at either side of the Mehrab wall of the southwest hall. The corridors are separated from the 6.15 m high transepts by 2.90 m high elliptical arcades. The simple altar (Mehrab), with the dimensions of 2.9 by 1.8 m, was added to the southern part of the temple. An arched opening, 7 m east of the main entrance,

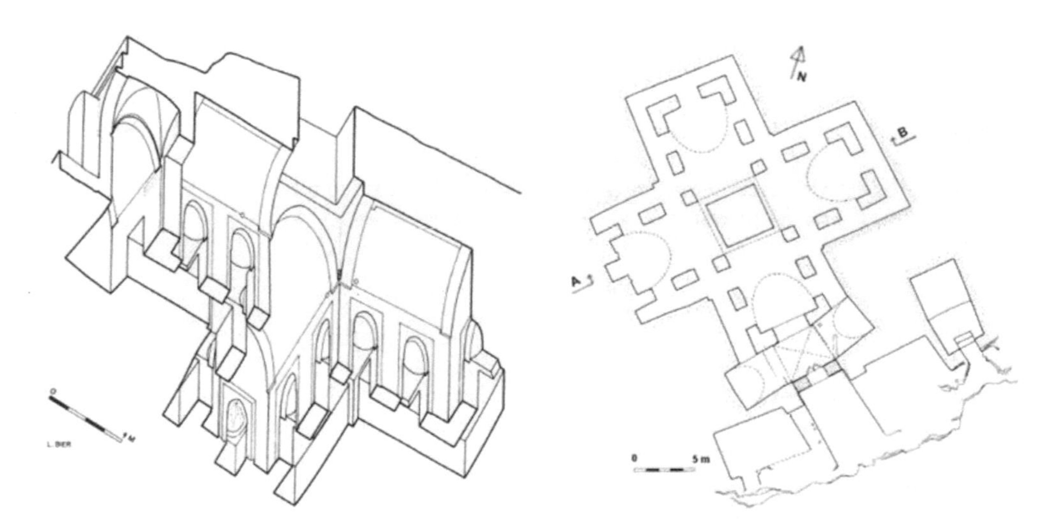

Figure 2. The axonometric projection and plan view of the Stone Mosque of Darab (Bier 1986).

Figure 3. The interior of the Stone Mosque.

gives access to a sunken rock-cut chamber with a flat ceiling measuring 3.10 m wide, 8.00 m deep and 2.95 m high that was probably the living space of the sentinel.

2.2 *Mehr temple in Maraghe*

The Mehr temple is located 6 km southeast of Maraghe city in the village of Varjuvy in East-Azarbaijan province (ICHO 1977). The temple is cut into rock and has a large opening (Figures 4 and 5). The entrance of the temple to the very end of the altar measures approximately 38 m. The walls of the entrance corridor are made of natural rock. The width of this corridor varies between 4.70 and 7.20 m. At the end of the corridor a rectangular opening of a width of 1.80 m is cut into the rock which leads to the main hall at the center of the structure. Just before the entrance to the hall and on the left side of the corridor a small square shaped room with a dome shaped roof consisting of a central Louvre is cut into the rock. An opening at the north-east side of this room leads to a 10×10 m chamber with a flat roof and an octagonal column made of rock, with approximately 5 m diameter, at its center. This chamber was probably added to the temple during the Islamic period for residing purposes (Shekari Nayeri 2006).

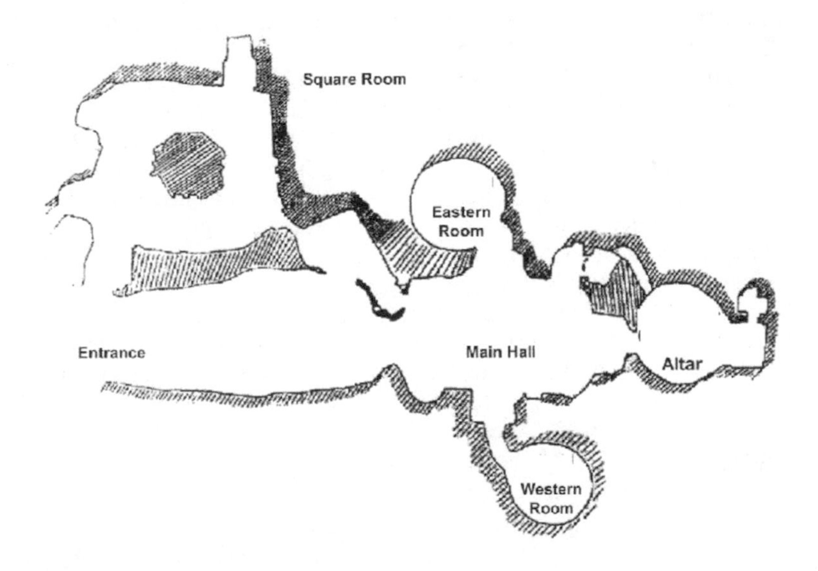

Figure 4. The plan of the Mehr Temple.

Figure 5. The main hall of the Mehr Temple.

Figure 6. The main space of the altar.

The main central hall has a length of approximately 12 m and a width of 6.30 m. Verses of the Quran are carved on four areas on the walls. On the eastern side of the hall is a corridor leading to a circular room and on its opposite side, on the western wall, a second fossa leads to a circular room with a diameter of 5.20 m, via a 3.10 m long and 1.30 m wide corridor. At the south-eastern side of the main hall a semi-circular corridor leads to the main altar of the temple. The altar can also be accessed from the southern part of the main hall via a 3.30 m long and 2.10 m wide platform (Varjavand 1972, Ball 1979). The altar is also circular with a diameter of 5.90 m. At the south of this chamber a rectangular niche of 2.60 m length and 2.30 m width is cut into the wall (Figure 6). In the past this chamber had a rock dome that has collapsed over the years and is replaced by a masonry dome. In the center of the altar chamber a grave of the Islamic period is found which is known as the grave of Mullah Masoum Maraghei who was one of the well-known scholars of the eighteenth century of the region.

Considering the plan, the structure and the types of rooms and chambers as well as the altar, it is thought that the temple was built by followers of Mithra for religious ceremonies. The arch shaped vaults and domes are architectural features of the Arsacid era. All three circular rooms have dome shaped roofs with a central Louvre.

During the post-Islamic period the temple was turned into a mosque probably during the fourteenth century A.D. Verses of the Koran were carved onto the walls and Islamic style ornaments were added to the domes.

3 TOMBS

Funeral practices are part of the culture of each nation. Different beliefs and practices developed throughout history as each tribe or nation has had a different perspective on this matter. Many of the ceremonies regarding the deceased relates to purification and sanctification. Some believed that bodies were impure and buried or cremated them quickly. Following the belief of life after death, fire and grave were looked at as a gate to the world after death. The variety of burial ceremonies is also evidence of different religious beliefs. Burying bodies in coffins, urns, or rock tombs were common in ancient times (Hashemi 2013). In those times the position of the sun in the sky was of importance during burial ceremonies and the graves faced to the direction of the sun. A guiding principle in Zoroastrian funerary is to prevent rotting flesh from coming into contact with soil, water, and fire to avoid polluting these elements. In Zoroastrian "Dakhmas" or "Towers of Silence" the majority of the flesh of the deceased is consumed by birds and the rest disintegrates through the action of sunlight and heat. The bleached and dried bones are then placed in an ossuary. An ossuary is a place to collect the bones of the dead. Often an ossuary is thought of as a box, similar to an ash urn, but it can also be a pit or a cave carved into rock.

As mentioned before Mithra (divinity of the Sun) and Anahita (divinity of the waters) were of special importance in ancient times for Iranians who constructed their temples or cemeteries towards the sun or flowing waters which in addition to its own purity also reflects the sunlight. Hence many Ostudans were cut into cliffs near rivers. Some of the other ancient underground spaces found in various parts of Iran are chamber-like manmade structures that had been excavated or carved into cliffs and served as tombs or graves in ancient times, following the Zoroastrian belief. Some of these structures are very simple and some are well decorated.

Most existing tombs in Iran are related to Median and post-Median era, and no other tombs have been identified (Ghirshman 1964). This is whilst in the land dominated by the Urartu kingdom (860 B.C. to 590 B.C.) especially around the Van Lake (in today's Turkey) many rock tombs were cut into the mountains for the kings. Therefore it could be concluded that rock-cut tombs might relate to Urartian art and architecture rather than having Assyrian or Median roots. Methods for excavating the tombs has varied in different eras and can be classified into the flowing three main groups:

a) Median era (678 B.C. to 549 B.C.)

The tombs of the Median era can be defined as the advanced art of Urartu (a tribe living around the Lake Van in today's Turkey before the Aryans). In contrast to the tombs of Urartu era these tombs contain stone carvings and ornaments in form of columns or half columns, and depending on the importance of the buried person, special carvings and facades may be found at the entrance of the tombs. Opposite to Achaemenid tombs where the owners are known, one cannot ascribe the Median to certain persons. As many of the tombs were robbed during history, it is not possible to make exact assumptions about burial ceremonies or the items that were placed in the graves.

b) Achaemenid era (550 B.C. to 330 B.C.)

Most tombs of this era follow a similar shape pattern and have a cruciform. The entrance of these tombs is usually at the intersection of the two cross arms. The upper part of the cross contains a fine carving and the lower part is left bare. Most tombs of Achaemenid Kings can be found in Naghsh-e-Rostam with the first being the tomb of Darius I. Succeeding kings followed his example for constructing their tombs.

c) Post-Achaemenid era

After Alexander's attack on Persia and the rise of Seleucidian and Arsacid dynasties, the role of the ancient religion became less important. The Seleucides were Greeks and had a different religion to Iranians. The Parthians also had a different religion and buried the dead differently. Hence, tombs with shapes similar to that of Medians and Achaemenians were not found anymore. Tombs were only built for Zoroastrians in soil or in rock, and the graves covered with stones or lime.

With the rise of the Sassanid, the Zoroastrian religion re-took its value but tombs of the Achaemenid form were never built after that period, however, public tombs continued to be constructed even after the attack of Arabs to Iran. Several examples of ancient tombs in Iran are briefly reviewed in the following text.

3.1 *Sahne Tomb*

These two tombs were carved in the cliffs of Shough-Ali Mountain, north of Sahneh city in Kermanshah province (ICHO 1932a). Both tombs overlook a valley and the Darband River but their facades face different directions. The larger tomb known as the tomb of Keikavous is also called the tomb of Farhad-va-Shirin or Shirin Gora (Kurdish for Large Shirin) by the locals (Figure 7). The southward entrance was excavated 12 m above ground and 50 m above the river, inside the mountain. In front of the tomb a rectangular porch is excavated with the dimensions of 5.82 × 1.75 m with the shape of a flying sun carved above it into the cliff. Two rock columns with a diameter of 55 cm were cut on both sides of the porch of which only the plinths have remained. The entrance behind the porch has a width of 1.32 m and a height of 1.70 m and leads to a chamber with the dimensions of 3×3.18 m and 2.45 m height. In the eastern and western sides of the chamber, two benches with the dimensions of 2.30×1.05 m and 50 cm height are cut out of the rock, within which the graves of 2.15 m length and 82 cm width are carved. Between these two

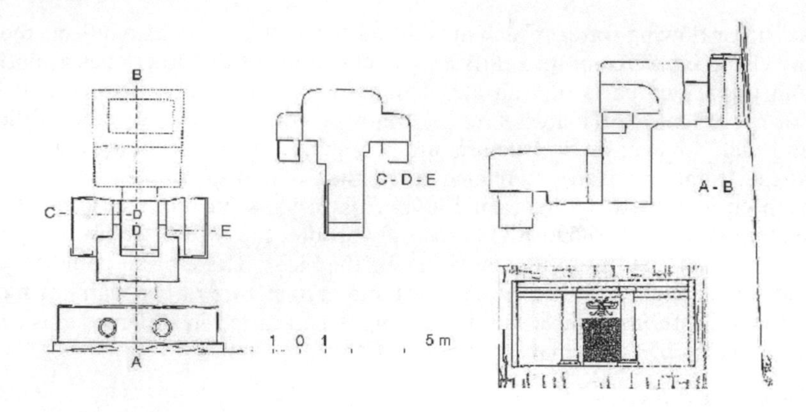

Figure 7. Plan view, front view, cross section and profile of Farhad-va-Shirin tomb (Von Gall 1966).

benches, a gangway with the dimensions of 1.90×1.45 m, leads to a room in the lower level. In the middle of this room the main grave is cut into the rock with the dimensions of 2.42×1.32 m which belonged to the owner of the tomb. Two other graves belonging to his close relatives are on both sides. The roof of this room is inclined. At the northern side of this room two small ledges were cut into the rock. Considering the style of carvings and form of the columns some researchers suggested that the tomb was built during the Median era (Ghirshman 1964, Herzfeld 1941). Others however believe it that it belongs to the Achaemenid era (Von Gall 1966, Vanden Berghe 1959). The smaller tomb which has an arc shaped roof is located about 100 m facing eastward and is known as Shirin Buchke (Kurdish Little Shirin). Its entrance is 1.25 m long and 80 cm wide. The tomb chamber has the dimensions of 2.5×1.65 m and a height of 1.10 m. The similar style of carving suggests that both tombs belong to the same era.

3.2 *Dokkan-e Davoud tomb*

This rock-cut tomb is located 3 km southeast of the city of Sar-e-Pol-e-Zahab in the province of Kermanshah (ICHO 1932b). This tomb which is positioned 10 m above the foot of the cliff was first discovered by Henry C. Rawlinson in 1836 (Rawlinson 1839). Dokkan-e Davoud was geographically identified as Median by Ernst Herzfeld (Herzfeld, 1941)

The southward tomb consists of a 1.95 m deep and 2.60 m high antechamber which is 9.60 m wide at the double frame of the entrance and 7.32 m wide at the back. Of the two columns in the antechamber only the bases and the capitals, of abacus form, are preserved. The bases are of simple shape, with plinths 0.83 m^2 topped by remains of round parts.

In the middle of the back wall a door (1.50 m high, 1 m wide) leads into a rectangular, barrel-vaulted tomb chamber (2.31 m deep, 2.83 m wide, 2.18 m high). Five small niches probably intended for lamps are carved into the chamber walls. On the left side of this chamber a cavity like a trough extends the full depth of the room; its floor is 70 cm lower than that of the chamber. This cavity is the sole provision for a burial in the tomb (Figure 8).

3.3 *Naqsh-e Rostam*

Naqsh-e-Rostam is one of the most beautiful and important historic sites of ancient Iran. It is about 4 km away from Persepolis in the northern part of Fars province in the Haji-Abad Mountain which in the past was known as the two-peaked mountain and is currently also called Hossein Mountain (ICHO 1931). Its construction has turned the southern face of the mountain to a surreal picture with a height of 64 m and width of 200 m (Figure 9). To its right another wall with a width of 30 m exists. The entire area contains memorials and reminders of the Elamite, Achaemenid, and Sassanid eras. It contains the tombs of several Achaemenian kings including Darius the Great and Xerxes I, rock reliefs of the Sassanid era such as the

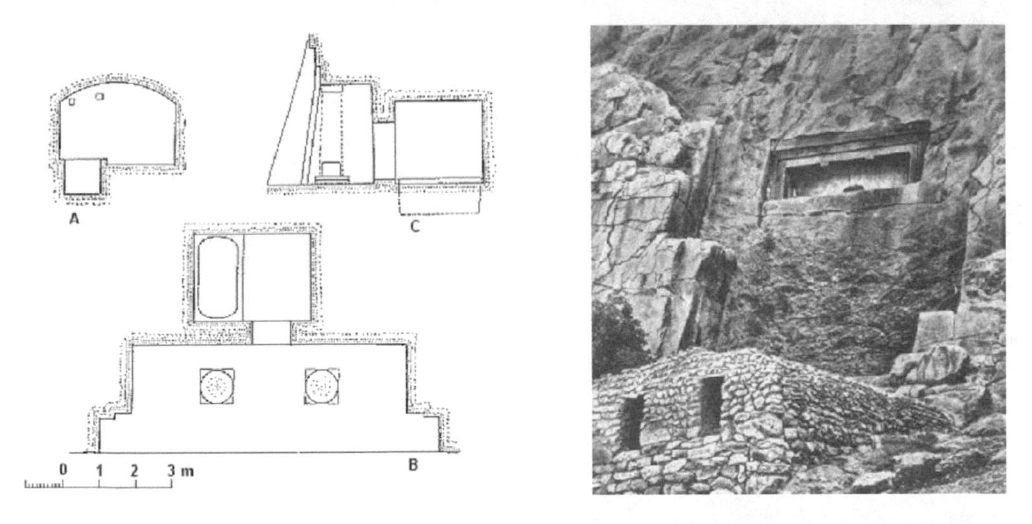

Figure 8. A. Internal section, B. Plan view, C. Cross section of Dokkan-e Davoud (Von Gall 1974, Ghirshman 1964).

Figure 9. Naqsh-e-Rostam (The cross wall).

coronation of Artaxerxes and the victory of Shapour over the Roman emperor, and a structure known as the Cube of Zoroaster.

At Naqsh-e Rostam which is also referred to as Necropolis, four tombs belonging to Achaemenid kings are carved out of the rock face (Schmidt 1970). They are all at a considerable height above the ground. The tombs are known locally as the 'Persian crosses', after the shape of the facades of the tombs.

One of the tombs is explicitly identified by an accompanying inscription to be the tomb of Darius the Great (521-485 B.C.). To its right the tomb of Xerxes I (485-465 B.C.) is located which is similar to that of Darius I except the absence of an inscription. It is the best preserved of the four tombs. The tomb of Artaxerxes I (465-424 B.C.) is located to the left of the tomb of Darius I. The tomb of Darius II (424-405 B.C.) is positioned at the west end of Naqsh-e-Rostam. Apart from slight differences the tomb facades are similar. The huge cruciform cavities are divided into three registers. The reliefs adorning the main panel of the top register show the symbols of Ahura Mazda, the sun, and an altar where the sacred fire is burning, beside the achaemenian king whose right hand is raised in prayer and holds a bow in the other. The middle register was sculptured to imitate the form of a palace and consists of four columns. The length of the palace front on the tomb is 18.57 m, and its height 7.63 m. The entrance to each tomb is at the center of each cross, which opens onto to a small chamber, where the king lay in a sarcophagus. These entrances are square shaped with a width of 1.75 m and height of 1.45 m. In ancient times these would be blocked using large bedrock slabs.

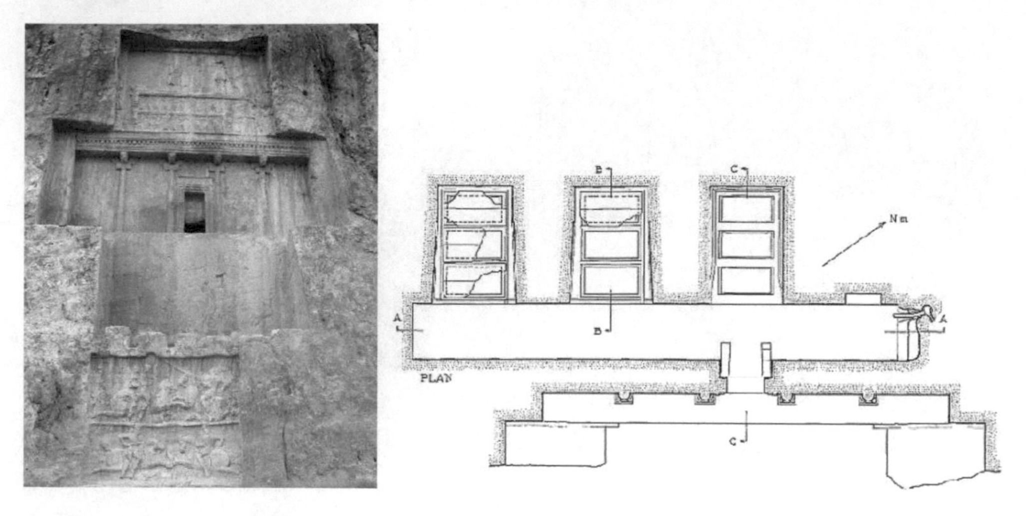

Figure 10. The tomb of Darius I and the equestrian relief of Bahram II in battle at Naqsh-e Rostam (Schmidt 1970).

As mentioned before the shape of the tombs are similar to each other except from the tomb of Darius which bears a trilingual cuneiform inscription. In this inscription Darius praises Ahura Mazda (the Avestan name for a divinity of the old Iranian religion, the Zoroastrianism) and counts his victories and expresses his thoughts.

The distance from the ground to the foot of the façade of Darius' tomb was about 15 m. The height of the façade is 22.93 m and the distance from the top of the façade to the top of the cliff is about 26 meters. The top and bottom register of Darius' tomb façade are equal in width 10.90 m. The height of the top register is 8.50 m and the bottom register 6.80. These two were probably intended to be of the same height (Schmidt, 1970). The vestibule of the tomb of Darius I is a neatly carved rectangular hallway with flat ceiling. It is 18.72 m long, 2.13 m wide and 3.70 m high. Its long axis is exactly parallel to the façade. Three vaults, each provided with three burial cists, pierce the rear wall of the vestibule (Figure 10). The nine graves belonged to Darius the Great, the Queen and few of his close relatives. The inner length of the cists is 1.92 m and their width 98 cm. Each grave was being covered with a lid made of rock slab. A 10 cm wide and 3 cm deep channel runs parallel to the vault walls in the floor of the vault that empties through oval holes into the vestibule, protects the cists against condensation or seepage. Almost 350 m^3 of rock had to be excavated with chisels and hatchets for the preparation of Darius' tomb. It is estimated that it took about 10 years to be built (Sami 1971).

4 CONCLUSION

Social memory is conveyed in ritual acts and studying such rituals is very interesting in that they shape and orient social practices. Understanding the backgrounds of such rituals can help for a better understanding of cultural evolvement and traditions. Everyday life and rituals are framed and defined in spaces inhabited by man. Their design and architecture expresses the sacral rules or specific needs that were translated into the structure that could serve as a setting for a ceremony or specific purpose.

Religious architecture is concerned with the design of places of worship such as temples, churches and mosques. The use of cruciform designs, vaulted halls, altars or Mehrabs are examples of architectural elements used in these structures.

A separate architectural typology is used for the purpose of burial rituals such as Towers of Silence (traditionally used by Zoroastrians which can be found in central parts of Iran) or Tombs that can be found in northern and western parts of Iran.

The examples reviewed in this paper are only few of many ancient underground structures and archeological sites in Iran, but should give an insight into the historical background of traditions and culture that shaped a big part of human history.

REFERENCES

Ball, W. 1979. "The Imamzadeh Ma`sum at Vardjovi: A Rock-Cut Ilkhanid Complex near Maragheh" In: *Archäologische Mitteilungen aus Iran (AMI)*, Vol. 12, pp. 329-340.

Ball, W. 1986. "Some Rock-Cut Monuments in Southern Iran" In: *Iran (Journal of Persian Studies)*, Vol. 24, pp. 95-115, published by British Institute of Persian Studies.

Bier, L. 1986. "The Masjid-I Sang near Darab and the Mosque of Shahr-I Ij: Rock-cut architecture of the Il-khanid period" Iran, Vol. 24, British Institute of Persian Studies, pp 117-130.

Cumont F. 1903. *The Mysteries of Mithra*, Nabu Press.

Ghirshman, R. 1964. *Iran – Proto Iranier Meder und Achämeniden*, Verlag C.H.Beck, München.

Hashemi, S. 2013. *The Magnificence of Civilization in Depths of Ground (A* Review *of Underground Structures in Iran – Past to Present)*, Shadrang Printing and Publishing Co., Tehran.

Herzfeld, E. 1941. *Iran in the Ancient East*; Archaeological Studies Presented in the Lowell Lectures at Boston. London, Oxford University Press.

ICHO (Iran's Cultural Heritage Organization) Records. 1931. Registration of Naksh-i Rustam under the No. 21 as a cultural heritage site. (In Farsi)

ICHO (Iran's Cultural Heritage Organization) Records. 1932a. Registration of the Rock Tomb (in Sahne, between Bistun and Kangavar) under the No. 148 as a cultural heritage site. (In Farsi)

ICHO (Iran's Cultural Heritage Organization) Records. 1932b. Registration of Dokkan-e Davoud Tomb under the No. 152 as a cultural heritage site. (In Farsi)

ICHO (Iran's Cultural Heritage Organization) Records. 1935. Registration of Masjid-e Sang (Stone Mosque) of Darab under the No. 229 as a cultural heritage site. (In Farsi).

ICHO (Iran's Cultural Heritage Organization) Records. 1977. Registration of the Mehri Temple under the No. 1556 as a cultural heritage site. (In Farsi)

Monneret de Villard, U. 1936. "The Fire Temples," *Bulletin of the American Institute for Persian Art and Archeology* V, 4, pp. 175-184.

Pirnia, M.K. 1983. "The architecture of Mosques, a way towards the kingdom of heaven" *Quarterly Journal of Art (Honar)*, No. 3, pp. 136-151. (In Farsi).

Rawlinson, H. C. 1839. "Notes on a March from Zohab… to Kirmanshah, in the Year 1836," *Journal of the Royal Geographical Society*, Vol. 9, pp. 26-116.

Sami, A. 1971. "The tomb of Darius the Great in Naqsh-e Rostam", *Historical Studies of Iran*, No. 37, pp. 107-136. (In Farsi).

Schmidt E. F. 1970. *Persepolis III: The Royal Tombs and other Monuments*, Chicago - Oriental Institute Publications (OIP).

Seyed Younesi, M.D. 1965. "Mehrab or Mihrab" *Quarterly Journal of Faculty of Letters and Humanities of Tabriz University*, No. 71, pp. 423-450. (In Farsi).

Stein, A. 1936. "An Archeological Tour in the Ancient Persis", Iraq, Vol. 3, No. 2, pp. 111-225.

Shekari Nayeri, J. 2006. "The Iranian Mithraic Temple of Varjuvi in Maraghe", Journal of the Iranian Studies, No. 10, pp. 109-126. (In Farsi).

Vanden Berghe, L., 1959, *L'Archéologie de l'Iran ancien*, Leiden.

Varjavand, P. 1972. "The Mehri Temple", *Historical Studies of Iran*, No. 42, pp. 89-100. (In Farsi).

Von Gall, H. 1966. "Zu den "Medischen" Felsgräbern in Nordwest Iran und Iraqi Kurdistan", *Archäologischer Anzeiger*, pp 19-43.

Von Gall, H. 1974. "Neue Beobachtungen zu den sog. medischen Felsgräbern" in *Proceedings of the 2nd Annual Symposium on Archaeological Research in Iran 1973*, Tehran, 1974, pp. 139-154.

White, L.M. 1990. *Building God's House in the Roman World: Architectural Adaptation among Pagans Jews, and Christians* (American Schools of Oriental Research Library of Biblical and Near Eastern archaeology), Baltimore and London: The Johns Hopkins University Press.

Wilber, D. 1955. *The Architecture of Islamic Iran: The Ilkhanid Period*, Princeton University Press.

Tunnels and Underground Cities: Engineering and Innovation meet Archaeology, Architecture and Art, Volume 1: Archeology, Architecture and Art in underground construction – Peila, Viggiani & Celestino (Eds)
 ISBN 978-0-367-46574-2

The Albinian way of design at the Milan Metro

Y. Kutkan-Öztürk
Middle East Technical University, Ankara, Turkey

ABSTRACT: Milan's underground transportation network was planned as an alternative public transport system for the 19th-century tram-based surface transportation, where its stations were designed by Studio Albini in collaboration with Bob Noorda/Unimark. Today, therefore, the system is assumed to be an example of 'Rationalism' in Milan, and as well as Italy.

Regarding this, the Albinian way of design that is still effective in the Milan Metro has affected the individual and collective behavior of its users in addition to the architecture, planning, social and economic development of the city from the material to symbolic value.

This paper aims to examine the architectural identity of the Milan Metro and Franco Albini's Rationalist approach to the interior design of the stations of the Red Line (Linea M1) along with Bob Noorda's well-known graphic designs of the signage system which would later influence the underground transportation networks of New York, Sao Paolo, and Naples.

On 1 November 2014, the 50th anniversary of the Milan Metro was celebrated with great enthusiasm. Since the inauguration in 1964, the success of Albini's innovations has made the system a symbol of modernity in the urban transport sector and thus gradually spread the architectural and functional standards throughout the world.

The Milan Metro is characterized by its strong uniformity in organization, materials, and design that can be clearly identified with the visual and technical coordination of all the details. In order to create a continuously uniform environment throughout the extensions of the network, Albini developed a strong architectural identity with specially designed furnishings and fittings in the existing structure accompanied by Bob Noorda's graphical illustrations that would later be the reference for the signage design of the New York Subway.

Since the system was designed as an equally dignified public place on the surface, the entire network with its innovative spatial design and structural construction is regarded as one of the most comprehensive urban design interventions in Italy.

1 HISTORICAL BACKGROUND

Due to its geographical position, Milan had been greatly influenced by the industrialization and socio-economic developments of the 20th-century and therefore became the economic, industrial and financial capital of Italy. As a result of the rapid and uncontrolled expansion of the city, the public transportation became a concern for local authorities. Consequently, the first projects to build an underground network were proposed as an alternative public transport system for the existing tram-based surface transportation.

In 1926, a national competition was organized for the design of the underground transportation network of Milan, and a comprehensive futuristic schemed project was awarded to be built by architect Piero Portaluppi and engineer Marco Semenza. However, the construction was hindered by bureaucratic regulations and economical inadequacy. Then in 1933, engineer

Semenza was commissioned to design 'the future Milan's metropolitan lines' which would later be called the baseline of the Linea M1. But this time, the World War II blocked the actions, and construction works eventually began to be implemented during the economic boom. (Kutkan, 2010)

In 1952, the City Council of Milan decided to appoint a professional to draw up the construction project of the metropolitan network. Thus, the Superior Council of Public Affairs commissioned engineer Amerigo Belloni to reorganize the project that Semenza had designed before. (Mai, 2005) According to Belloni's design, the entire network consisted of four independent main lines, each identified by a number and a color.

Linea M1 – *red*;	12.5km from Lotto to Sesto San Giovanni
Linea M2 – *green*;	8.5km from Piazzale XXIV Maggio to Piazza Piola and Via Pacini
Linea M3 – *yellow*;	5.5km from Via Andrea Solari to Via Lunigiana passing through the Centrale Station
Linea M4 – *blue*;	8.0km from Piazza Medaglie d'Oro to Piazza Firenze (Busato, 2005)

After all those previous attempts, the Board of Public Affairs approved the project in 1954 and finally, in 1957, the excavation works of the Milan Metro began with cut & cover system, a considerably different method that would later be called as 'the Milanese technique'.

In this open-air excavation system, where the height of the tunnel is limited, construction is initiated by forming two sides of the gallery walls and bentonite is applied over these surfaces to obtain a robust bulkhead structure under the ground. Subsequently, reinforcements are used to strengthen the walls and the structure is finalized by applying concrete over the constructions.

Throughout the following years, the construction works continued into the inner parts of the city and in 1960, the Piazzale Duomo station was built to be a connection point to the third line in the future. After completion of the structural construction, the newly elected Municipal Council commissioned architect Franco Albini and his partner architect Franca Helg to design the stations of the first line, Linea M1, of the Milan Metro in 1962.

2 ARCHITECTURAL CHARACTERISTICS

The Milan Metro is a considerably different system from most of its counterparts with its network very close to the surface, in which the relatively low ceiling height became its structural characteristic. In Linea M1, all the stations are identical except for the depth difference of the train tunnels and are built on two main platforms only separated by structural partitions.

The structure of a station consists of two main levels, the actual platforms where the trains pass and the mezzanine level of access, control, and public services. (Spinelli, 2006) The large mezzanine, namely concourse, is the starting point of all access passages that reach the entrance and exit directions on the surface. Each concourse is designed to contain at least one newspaper and tobacco bar with further commercial activities in the main stations and is divided into two by the controller's kiosk and the turnstile line. Both levels can be viewed at once, therefore, the access stairs to the levels are placed on both sides, as the entrance and the exit, at different widths to facilitate the outflow of passengers arriving at the same time. (Albini & Helg, 1966)

Since the structural works of the stations had already been over, the finishing works on their distribution layout were carried out with very limited possibilities of intervention. For that reason, to obtain 'uniformity' in their design language, Albini focused on a search for innovative and reproducible materials that would be applied repeatedly in different varieties and combinations across all 21 stations.

The revolutionary cladding material 'silipol' - made of cement, binder, and grains of granite, marble, and quartzite, by Fulget - supported with dark brown painted iron plates, and anchored to the shear walls in different widths to provide electrical and mechanical installation channels. (Spinelli, 2006) On the other hand, the heights of the planes were determined as a constant in order to support the information ribbons. A color-coordinated band with a height of two meters along the walls defined the station name every five meters, while the second one displayed the exit and transfer directions and safety signs designed by Bob Noorda, who created custom fonts and icons derived from Helvetica in line with Albini's interior design. (Byrnes, 2016)

A plastic-based vinyl paint in dark brownish/greenish color was used to paint the ceiling, the columns and the other of the vertical surfaces to minimize irregularities on the exposed concrete while providing flexibility to withstand the vibration produced by the trains.

Figure 1. Caption of the stairways of Amendola-Fiera Station. ©Fondazione Franco Albini.

The fluorescent luminous channels were the indicators of the main routes and installed as the lighting network of the system highlighted the concourse level and the platforms as continuous strips along the access corridors. (Albini & Helg, 1966)

As the flooring material, pointed stamped black rubber tiles - studied and designed especially for the Milan Metro, by Linoleum subsidiary of Pirelli - were applied because of their noise reduction function, therefore helped to acquire the overall acoustic comfort in the stations. Granite tiles 'serizzo formazza' were used to cover the staircases that were equipped with electrical installment to avoid frost during winter.

Albini 'tied' the entire structure with a tubular element painted in red enamel as the most characteristic feature of the project, which serves both as the handrail and a partition of the spaces. It has a diameter of 47.5mm with standard curves welded in the terminal laps so, it was described as a strand of Ariadne that facilitates descents and ascents. He gave a sense of aesthetics to the functions as registration of tickets, flow management and passenger counting by his turnstile design that was made of metal and colored red as the rest of the architecture.

Albini exclusively worked on the typology of the clocks that had been accompanying the Milanese public transport throughout its history. He aimed to reflect the simplicity and functionality like the entire structure with its well-defined numbers, large notches and the three red hands which were placed in a painted metal case in the same dark color of the walls. (Anon, 2014)

In 1962, Albini and the famous graphic artist Bob Noorda came across to work on the signage of the Linea M1. Consequently, the system became another innovation in the 1960's Italian architecture and would be the reference for other metropolitan systems as New York, Sao Paolo, and Naples.

Figure 2. Caption of the platform level of San Babila Station. ©Fondazione Franco Albini.

3 THE ALBINIAN THEORY

"The Albini style is coherent, rigorous, almost maniacal in its attention to detail, but at the same time it always stimulates new emotions with surprising traces that mark the space." (Bucci, 2009)

3.1 *The rise of Rationalism*

During the first half of the 20th-century, Italy had still been under the effect of stylistic methods, and modernist ideas were not considered as the usual architectural concepts. After the change in government, architecture started to be seen as the instrument of political propaganda to show power and promote their ideological ideas. Therefore, in the 1930's, the Fascist regime began commissioning works from young architects practicing in international avant-garde styles who focused on form avoiding radical utilitarian functionality. In that period, aesthetic purity was regarded as the uppermost ideal that a person could aspire towards. This attitude was moderated by the impossibility of a complete negligence for utility in architecture, however it fostered a deep-seated notion of functionalism. (Di Robilant, 2018) Through the belated industrialization, 'modularity' and 'serial repetition' started to be implemented as architectural criteria, with an intention of carrying the building industry into success.

In the early years of his career, Albini witnessed the battles against political difficulties because of Italy's isolation from the avant-garde mainstream that fought for modern architecture. For that reason along with his inward character, he was shaped into a relatively suppressed personality among his contemporaries, such as Italian Rationalists and Bauhaus proteges. (Jones, 2016)

Throughout the years from the 1930's till '45, Albini made researches about vernacular architecture, which spontaneously affected the formalization of techniques and typologies of the Modern movement. He was influenced by his milieu that encountered with its flows of thought, in which certain essential aspects of the spirit of the places and the times might be counted as the actors of his method. (Helg, 1990)

In that sense, Albini's architectural point of view can be easily distinguished from the others by his effort to combine the poetic nature of life with the reality of social organization accompanied by a hidden opposition. Consequently, authorities began to see him as one of the pioneers of the Rationalism movement that implied a complete devotion to logical, functional and mathematically ordered architecture. With this new movement, architecture started to be considered as a new idea of space based on the actualization of material and its geometrical use in design rather than abstract concept along with intangible elements. (Cortesi, 2004) Yet, it did not represent a clear opposition to fascism where some Rationalist critiques claimed it as a potential instrument to identify Italian Modernism in its appropriate architectural state. (Leet, 1990) Thus, because of the confrontation of social and cultural lifestyles, Italian architecture acquired a transient meaning. (Sartoris, 1990)

From 1938, Albini and his fellow Rationalists began to deal with public and green spacc, transportation, and infrastructure into new collective compounds through urban planning. Unlike his colleagues, Albini saw Modernity as a powerful 'interior' idea in which 'the room' was the most essential unit element. Therefore, he used this notion as the relation of the room to the building and the building to the city while responding to specific urban modifiers. (Jones, 2016)

3.2 *The Post-war Period and the Albinian Theory*

In the second half of the 20th-century, Rationalism began to be seen as a limitation for hindering Italian architecture and urbanism, and Neorationalism emerged with an intention to develop a city understanding beyond simple Functionalism. Within this new notion, 'the city' was assumed as architecture itself, but not as an image or the styles, it was the architecture as construction and therefore construction of the city continued over time. In this sense cities

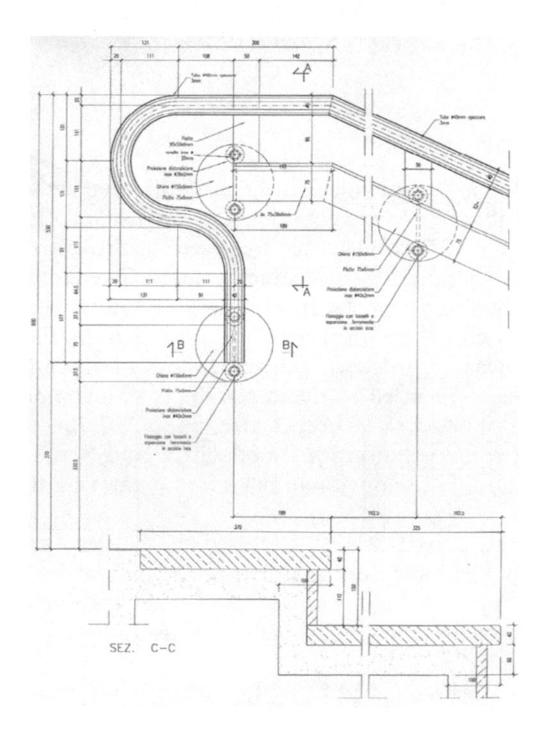

Figure 3. Caption of the project of 'il corrimano' of the Milan Metro. ©Fondazione Franco Albini.

Figure 4. Caption of the passageways of the Milan Metro. ©Fondazione Franco Albini.

acquired life, culture, environment, space and civilization those have already existed in the context of man. (Visentin, 2012)

Architecture was considered as a system of thought with mythical condition that a physical construction might only be achieved as a result of the ideational construction. (Peckham, 2007) In this sense, the priority was generally given to structural elements and the utilization of their components. Yet, construction elements were reduced to their minimal structural and aesthetic condition along with their rationally considered connections and details.

Albini usually used his emotional intelligence that was rigorous without being rigid or schematic and was restlessly open to new inquiries or new experiences, besides his technical knowledge. He accepted authenticity as an indispensable part of architecture and developed recognizable lines to characterize the post-war period, where 'anti-fascism' notion was the only hope.

As Albinian architecture was more significantly related to the cultural references of his epoch, Kafka might be identified as his rational soul-mate due to his aim of delivering products to posterity. (Tentori, 1965) He was obviously more concerned with abstraction and formal manipulation of essential elements than a didactic presentation of projects. (Leet, 1990) Therefore, critics tried to determine his approach as an open road or an intangible sense tended to a magical abstraction. Helg portrayed him as a person who had natural gifts of fantasy in order to invent new forms along with the control of proportions, the richness of composition that came easily to him. (Helg, 1990)

As being a protagonist of Italian Modernism, Albini was also known as a firm supporter of the current technology of his time in order to produce recognizable works for the upcoming generations. He defined this attitude as such that these generations would have to follow the innovations and know the new materials, as had their predecessors who were familiar with the old methods. He followed an entirely different method than his colleagues by using information and technology as the prerequisites before forming the logical context of his project. On the other hand, he had never been a technical designer, who aimed to implement a specific program rather than acting in accordance with spontaneous requests of the works. (Cortesi, 2004) He described this method as an imminent result and exemplified it *"just as it is impossible to walk without friction it is also impossible to design without conditioning factors. These factors should not be considered as constraints but rather incentives."* (Leet, 1990)

As an introspective designer, Albini focused on the essence of the design problem in order to provide simple and functional solutions for architecture at variable scales and functions. (Jones, 2016) He handled his projects with a great respect for the specific qualities and conditions of each problem and concentrated on a reductive approach to each design. For this reason, his rigorous minimalism demonstrates an innovative and profound work that can be achieved by restricting the boundaries of exploration.

For instance, in his studio, the formation phase of a project usually started with a pennant action that Albini usually sat on the table and began sketching according to the data obtained during the observation period. Then the operation initialized by breaking down the former idea into smaller elements and continued by using them as modules and frames in various organizations along with the details. This sequence might also be named 'the Milanese method' in which the technological reference of prefabrication was accepted for the composition since it was based on a logical process.

Albini's works play an essential role to portray and understand his period as being identified by their revealed construction methods, spatial simplicity, and the material innovation. He, above all, was a great master, then later an architect and a designer. Although his contributions to architecture are located within the borders surrounding the city of Milan, his influences continue to be widely recognized in other geographies. Therefore it is acknowledged that the Albinian way of design was local but international. (Jones, 2016)

4 CONCLUSION

On 1 November 1964, the Milan Metro started to work with a great opening at Lotto Station. In the same year, the intuitive and uniformly designed project by Albini, Helg, and Noorda, was rewarded with the Compasso d'Oro for its superior quality of architectural coordination and configuration of the signage system.

The Milan Metro stands out as a pioneering example not only for its innovative structural system but also for its creative and inspiring solutions in terms of its tectonic organization, materials, and design approach that can be seen in the details.

From the structural point of view, the entire system is regarded as one of the most comprehensive urban design interventions ever to be realized in Italy as the underground network was constructed very close to the ground level. Considering its structural layout, the station consisted of two levels where the mezzanine was the place of main public access and served as shopping galleries as well as hosting passageways and control facilities.

Since the structural system was already built, Albini conducted research on finishing works and he developed a strong architectural identity with his innovative and reproducible materials that still enhance the architectural value of the system in an extraordinary way. He also used specially designed furnishings and fittings in the existing structure to create strong uniformity throughout the extensions of the system which would later influence the underground transportation networks of New York, Sao Paolo, and Naples.

As a consequence, it would be a very erroneous act to frame Albini and his design approach to a very simplified way yet it is very important to understand the true meaning behind Albini and the Albinian way of design therefore further studies needs to be established in a more detailed way.

REFERENCES

Albini, F. & Helg, F. 1966. Immagini della Metropolitana di Milano Relazione Tecnica di Progettisti. *Domus*: 42–48.

Anon. 2014. MetroMilano50. *Fondazione Franco Albini*. http://www.metromilano50.com/.

Bucci, F. 2009. *La Scuola di Milano/The School of Milan*. A. Monestrioli, ed. Milano: Mondadori Electa S.p.A.

Busato, C. 2005. *La Cronistoria della Metropolitana di Milano*. Venice.

Byrnes, M. 2016. The Undervalued Simplicity of Bob Noorda's Vision for Milan's Metro. *Citylab*. https://www.citylab.com/design/2016/05/the-undervalued-simplicity-of-bob-noordas-vision-for-milans-metro/483782/.

Cortesi, A. 2004. L'architettura delle Connessioni - Franco Albini. *Dialoghi di Architettura*: 19–45.

Helg, F. 1990. Franco Albini - Architect and Teacher. In S. Leet, ed. *Franco Albini: Architecture and Design 1934–1977*. New York: Princeton Architectural Press: 13–18.
Jones, K.B. 2016. *Suspending Modernity: The Architecture of Franco Albini*. New York: Routledge an imprint of the Taylor & Francis Group.
Kutkan, Y. 2010. *Franco Albini and His Approach on Metropolitan of Milan with Franca Helg and Bob Noorda*. Politecnico di Milano.
Leet, S. 1990. Franco Albini and the Scrutiny of the Object. In S. Leet, ed. *Franco Albini: Architecture and Design 1934–1977*. New York: Princeton Architectural Press: 21–40.
Mai, M. 2005. Milan Underground, 40th Anniversary of the Opening of Line I. *Abitare*: 112–119.
Peckham, A. 2007. The Dichotomies of Rationalism in 20th-Century Italian Architecture. *Architectural Design*, 77(5): 10–15.
Di Robilant, M. 2018. The Aestheticization of Mechanical Systems: Gio Ponti's Montecatini Headquarters, Milan, 1936–39. *Journal of the Society of Architectural Historians*, 77/2(June): 186–203.
Sartoris, A. 1990. Franco Albini and Rationalism. In S. Leet, ed. *Franco Albini: Architecture and Design 1934–1977*. New York: Princeton Architectural Press: 45–46.
Spinelli, L. 2006. *The Places of Franco Albini - Itenaries of Architecture*. Milano: La Triennale di Milano and Mondadori Electa S.p.A.
Tentori, F. 1965. Opere Recenti dello Studio Albini-Helg. *Zodiac*.
Visentin, C. 2012. The Nocturnal Aesthetic of Italian Modern Architecture and Art from Post World War II to the 1970s. In E. Monin & N. Simonnot, eds. *Luminous Architecture in the 20th Century (1907–1977) Atti Simposio Internazionale 10–12 dicembre 2009*. Nantes: U. Snoeck: 150–157.

Tunnels and Underground Cities: Engineering and Innovation meet Archaeology, Architecture and Art, Volume 1: Archeology, Architecture and Art in underground construction – Peila, Viggiani & Celestino (Eds)
© 2020 Taylor & Francis Group, London, ISBN 978-0-367-46574-2

Mobilizing cultural resources: The functional role of heritage in metro projects

M. Laudato
Aarhus University, Denmark

ABSTRACT: The article provides for a broad conceptual framework about the relationship between urban Mass Rapid Transit projects and the growing presence of cultural heritage practices within these infrastructures. By adopting the theoretical setting of Critical Heritage Studies, the paper contests the idea that heritage constitutes merely an additional aesthetic element inside projects of mass urban transport, outlining rather its functional role in communicating political agendas by promoting and/or excluding identity representations in the urban landscape.

1 INTRODUCTION

The relationship between rapid transit systems and cultural heritage is a rather controversial one: on the one hand, it is something that is easily perceived in the experience of Metro spaces around the world. Getting on a city subway is always an experience bringing historical, literary and cinematic memories to the fore. Every subway has its own atmosphere, its character that makes it a unique part of every city identity. However, Metros, like railways, highways and airports, are also just parts of the transportation network, simply functional infrastructures for the mobilization of objects and bodies, preferably as fast and efficiently as possible.

Apparently, within these infrastructures, the presence of cultural heritage interventions would seem to constitute a sort of marginal and accessory extra in the general budget.

But yet, the concrete and steel spaces of Metros are crowded with artistic and architectural interventions, archaeological and historical exhibitions, literary, cinematographic and musical references. In the underground Metro space, cultural heritage is often widely and explicitly exhibited, promoted, collected, performed and somehow also contested.

The purpose of this paper is to investigate the conceptual foundations for which the relationship between cultural heritage and rapid transit systems has developed and constitutes nowadays a practice increasingly integrated into urban mobility projects.

The observations collected in this article have been developed in the elaboration of a thesis for the Master in Sustainable Heritage Management at Aarhus University, Denmark. Given the absence of specific study cases and in-depth analysis of empirical data, this article is here proposed as a contribution to the conceptual definition of the topic.

We will define the topic through a general overview of the phenomenon and through the definition of the main socio-cultural implications of heritage presence in Metro systems.

After a brief description of the current state of research, we will proceed with the description of the many different forms of cultural heritage expression within Metro infrastructures. On the basis of these observations we will frame the theme of the relationship between heritage and Metro through the definition of some main involved concepts, such as the concept of "sense of place", the social divide implied by the concept of "mobility capital", as well the presence of "intangible goals" in urban mobility programs.

1.1 *Metro infrastructures and urban identity*

The topic of the role of underground Rapid Transit systems in the urban identity negotiation process has been examined mainly in the field of cultural geography, where transportation infrastructures has been analyzed as "windows in social worlds" (Angelo and Hentschel 2015), or as technological infrastructures for the creation of individual and collective subjectivities, as in the case of character of the "*New York commuter*" (Höhne 2015).

Regarding the topic of the role played by Metro structures as *constructions and constructors of sociocultural meanings and urban identities*, has recently emerged a convergence of research coming from different disciplines, such as historical geography, transport planning theory and public policy studies (Merrill 2015, p. 78).

In Samuel Merrill's research about Berlin, the U-Bahn has been analyzed as an ideological and political branding tool for the city after the 1990 reunification.

Merrill analyzes the historicizing discourses put in place by the German public authorities, both local and national, in order to re-brand the Berlin U-Bahn as a metaforic reunification factor of the city, after 1989.

1.2 *Defining heritage*

It is impossible to reduce the definition of heritage to a formula, as it is, by its very nature, an unstable and contested idea (Davison 2008, p. 40). It is also a concept which has rapidly changed over time, and naturally subject to variations of meaning in different cultural contexts.

It is therefore necessary to dwell on the question of whether the forms of cultural sociocultural expression in the Metro spaces can be properly defined as heritage forms.

The term heritage, in its semantic root, still maintains the stratified meaning of object or cultural practice "inherited" from the past, endowed with an intrinsic value of historical testimony.

However, several contemporary authors have stressed that the historical-temporal dimension of the term has been strongly eroded in the postmodern era. David Lowenthal emphasized how the relationship between heritage and past is actually a relationship that has little to do with the actual historical value of a heritage object or a cultural practice. Heritage, in the definition of Lowenthal, *is not a testable or even reasonably plausible account of the past, but a declaration of faith in the past* (Lowenthal 1996, p. 121).

By limiting the absolute predominance of historical value in the definition of what heritage is, Graeme Davison points out that *heritage, what we value in the past, is defined largely in terms of what we value or repudiate in the present or fear in the future* (Davison 2008, p. 33).

Davison's definition highlights the importance of the process of assigning "value" to heritage resources rather than their "implied" relevance as historical testimonies.

Several authors have stressed that the definition of heritage is closely linked to that of the values to which its physical and ideal manifestations are associated: values of aesthetic and representative character.

In this perspective, what makes a sociocultural form of expression a heritage form are essentially the values, *intended as attributes given to sites, objects and resources, and associated intellectual and emotional connections that make them important and define their significance for a person, group or community* (Jameson 2008, p. 57).

Therefore, in a "value-based" approach, the fact that an object, space, or practice constitutes heritage resides basically in the act of recognizing and attributing value attributes in it. It is a relativistic view of the term, dependent essentially by the contingency of the recognition process by the purposes of the subjects acting in the process.

2 FORMS OF HERITAGE IN METRO SYSTEMS

The interaction between heritage and Metro infrastructures may occur under many different forms, as well as heritage in Metro spaces can perform many different functions, daily experienced by millions of people. We will briefly explore the main forms and functions assumed by heritage in rapid transit spaces.

2.1 *Architecture and infrastructure*

The modern idea of creating underground spaces in order to fulfill mobility functions had, since the very beginning, to deal with the "unnaturalness" of the new environment and the condition of estrangement and alienation to which the commuters would be subject in it.

A clear point was that the functional colonization of the underground dimension needed a programmatic aesthetic mediation, through architectural solutions.

The architect of London's Underground Charles Holden featured the stations design to simple and functional modernist forms, by adopting the linear design of the Art Deco (Wolmar 2012).

In the years following the First World War, the reduction of transport costs through massive public investment provided an opportunity for the theory of "transport democratization", giving mobility to a larger part of urban population, connecting the suburbs to the city center and to the production areas.

The nationalization of public transport and the investment of large financial resources, coincided with a progressive monumentalization of urban underground spaces. The most striking case is probably that of the "Lenin" Moscow Metro, where the stations were built and furnished as "Underground Socialist Cathedrals" (Hill 2015).

After the Second World War, the urban transport model based on rapid transit systems went through a critical phase, as a result of the industrial impulse given to the mechanization of private transport, based on the large scale diffusion of low cost cars.

After the oil crisis of 1973, the issue of mass public transportation closely linked to that of environmental sustainability, giving new ideological and programmatic impetus to the economic investment in this sector.

In response to the energy crisis, but also to the increasing pollution of urban environment, about 150 Underground Rapid Transit systems have been built in the past 40 years, among them approximately 200 currently active.

Under the principles of the so-called "New Architecture", subways structures have been built with the programmatic intent to provide environmentally sustainable transport systems, as well as aesthetically and socially pleasant places.

As Cervero outlines, the New Architecture Metro represents a built form and a mobility environment where transit and the built environment harmoniously coexist, reinforcing and enhancing each other in the process (Cervero 1998, p. 4).

The exceptional aesthetic quality of modern subways clearly transcends the pure functionality of underground transport infrastructures. For Cervero and Calthorpe, the underground rail is today a "planning tool". *For an architect rail is an opportunity for design a stunning interior; for a civil rights leader it is a chance to correct past discrimination; and for an ordinary rider it may simply be a fact of hometown pride* (Cervero 1998, p. 10).

In this sense, the architecture of the modern subways aims to be an integral part of the urban heritage as well an integrated part of urban identity.

2.2 *Archaological and historical exhibitions*

In contemporary urban planning, archaeologists are statutorily obliged to overseeing how past materialisations of cities are excavated, recorded and preserved (Novaković 2016, p. 8).

Metro construction sites are no exception to this normative practice, and in many cities underground stations have become one of the privileged places, outside the institutional spaces of museums, where archeology is preserved and exposed to the public.

After the cases of Rome (1955) and Mexico City (1969), the practice of preserve and exhibit *in situ* archaeological structures and artifacts has become increasingly widespread. The leading case has been the pioneering and scenographic set up of Athens Metro (1994), where many stations constitute full museum exhibitions (Parlama, Stampolidis 2001).

In the last 20 years many archaeological exhibits have been created inside Metro stations, as in the case of Madrid (2008), Naples (2007), Oporto (2004), Vienna (1994), Paris (1989), Lyon (1978) and Amsterdam (1980).

Thematic "archaeological stations" have recently been set up in the Sofia Metro (2016) and in Rome (2017) and Amsterdam (2018), while others are under construction in Algiers (2018), Istanbul (2019), Thessaloniki (2020) and Cologne (2023).

The more Metro systems are built, more the presence of archeological heritage in Metro spaces becomes a widespread and consolidated phenomenon.

The location of historical or archaeological heritage materials within the Metro stations represent a very strong form of ideological representation and promotion of the link between the new artificial transport infrastructure and the stratified historical memory of places.

The issue of preserving archaeological artifacts "in situ" or in "re-contextualized" conditions is one of the recurring themes in the scientific and academic archaeological debate. The basic concept that supports every critical position, in favor or against the conservation in situ of historical-archaeological evidences, is basically that of preserving authenticity.

Authenticity has been approached and treated in many different ways in archaeology.

Cornelius Holtorf states that objects are not authentic in themselves, but are *made authentic through particular, contextual conditions and processes taking place in the present.* (Holtorf 2005 p. 117).

The topic of authenticity does not, however, only represent a topic of academic relevance. Authenticity is also tied to topics of power and authority. Holtorf outlines how archaeologists invoke an authoritarian role in defining what is authentic and worth of protecting, and what is not.

Laurajane Smith, claims the fact that *Archaeological heritage management embodies a process of cultural domination, and that archaeology is used within State discourse to arbitrate on cultural, social and historical identities* (Smith 2008, p. 63).

The presence of archaeological heritage in the Metro spaces therefore presents a number of different aspects.

In one hand this presence can be interpreted as a step in the direction of the transformation of the "non place" of transport infrastructure, in a place integrated with the historical context of the stratified urban identity. Archaeological heritage, by bringing in the Metro its authority, brings also its own potential of cultural significance, somehow empowering the Metro space as an integral part of the historical narrative underlying the local sense of place.

On the other hand, the archaeological heritage management practice represent a challenge to the technical and normative role of mobility authorities, through the introduction of authoritative discourses and practices, linked to the preservation of historical memory.

2.3 *Art collections*

Many different Underground Systems are hosting artworks and building up their own public functional identity as places where artistic heritage is acquired, preserved and promoted to the public as an integral part of the transport service.

Among the many examples we cite here Brussels Metro (1960), the first creating its own art collection, the Montreal Underground (1966) and Lisbon´s Metropolitano (1950), as well Vienna subway lines U1, U2 and U3 defined as "Art lines". Stockholm Metro is branding internationally itself as the "world's longest art gallery".

Originated usually from artistic personalities, Metro collections are mainly hosting artworks by local or national artists, and the themes portrayed in the artworks are often related to local culture and history. Metro collections are characterized as forms of public art, closely linked to the specificity of the places where they are produced and exhibited.

The functionality of public art in promoting the reshaping of local urban spaces for political and economic purposes through the visual reorganization of public spaces is not a new factor, dating back to the origin of public patronage.

In his seminal The *Production of Space* Henri Lefebvre argues that under modern capitalism, space has become more abstracted, homogenized and compartmentalized for the purposes of commercial exchange and political inclusions/exclusions (Lefebvre 1991).

In Lefebvre interpretation, art can be playing a key role in breaking the capitalistic commodification of space: a concentration of high-quality spaces, accompanied by architectural facilities of great impact and aesthetic value can create what is called "a surplus of art".

In conjunction with elements of art, architecture and urban planning, public space can significantly broaden the mind of the "creative class", the social class whose economic function is to form new ideas. In a way, public spaces inhabited by a "surplus of art" becomes an arena of new creative explorations, experiments or risk undertaken (Nadolny 2015, p. 31).

By introducing creative interventions, public art performs as a means to alleviate uniformity in the Metro design, because otherwise underground stations would be framed as non-places: but the question whether art can or should be place-representational, thought-provoking, beautiful, socially challenging, and decorative, presents certain challenges.

The predominance of institutional approach in the selection of artworks has often led to potentially contradictory definitions of the public meaning of art in Metro spaces.

As pointed out by Wendy Feuer, founder of the Art for Transit program in New York, *should a state agency that serves such a broad spectrum of the population sponsor work that is esoteric, confrontational or politically-or-sexually controversial? What is the responsibility to people who have not paid to see art but have paid to be provided with transportation services? Is the art doomed to mediocrity?* (Feuer 1989, p. 151).

2.4 *Spontaneous, unauthorized and dangerous forms of heritage*

The Metro space is governed by a strict set of behavioral rules, more or less explicited and legally defined, rules that go often far beyond simple precautional behavioral limits, designed to protect travelers' safety.

Michel Foucault in *Discipline and Punish* analyzes the many normative mechanisms that, under the form of regulations for the use of space, have surreptitiously reorganized the functioning of power. *"Miniscule" technical procedures acting on and with details, redistributing a discursive space in order to make it the means of a generalized "discipline"* (Foucault 1975).

In opposition to the Foucault's discipline strategies, expressed in the form of technical regulations of behavior, Michel De Certeau identifies some forms of anti-discipline, denominated "ways of operating". They constitute *clandestine forms taken by the dispersed, tactical, and make-shift creativity of groups or individuals already caught in the nets of discipline* (De Certau 1980, p. XVI).

Despite, or perhaps precisely because of the existence of a strict regulatory settings, the Metro provides a stage for the manifestation of what Dick Hebdige defines "Subcultures", *forms of subversion to normalcy*. Subcultures bring together like-minded individuals who feel neglected by societal standards and allow them to develop a sense of identity.

One of the most known subcultures inhabiting the Metro spaces is that of graffiti makers.

The appearance of graffiti in stations and subway trains originated in New York in the late '70s

The famous story of the NYC subway graffiti culture and the almost two-decade long struggle of the authorities to eradicate tagging represent the starting point of the controversial question, if graffiti are a form of vandalism or a form of art (Chalfant, Jenkins, 2014).

Graffiti advocates perceive this practice as a method of reclaiming public space or displaying an art form; their opponents regard it as an unwanted nuisance, or as expensive vandalism.

Another unauthorized form of artistic expression in the Metro is busking: musicians, dancers and actors running performances in change of donations.

The daily experience of buskers in the Metro, regardless of the artistic quality of their performances, takes place on a liminal ground, the space between the support and appreciation of the public, the indifference, the annoyance or the hostility of those considering to be molested, the benevolence or the persecution of the security staff in applying the regulations.

All over the world, without exception, public performances in the Metro spaces are prohibited by the regulations, regardless of the appreciation of the audience.

William Whyte pointed out that one of the most important qualities of music performance in the public realm is that it produces social activity and social space (Whyte 1979).

The image of harmless musicians arrested and expelled from the Metro by the police is perhaps one of the most noticeable manifestations of the normative power that, supported by his own regulations, establishes the degree of sociability considered acceptable or not, within the underground spaces.

It is still unclear if graffiti and busking can be considered forms of heritage, but it appears quite clear that the assimilation of these spontaneous forms of art into manifestations of local identity takes place through heritagization processes.

If heritagization constitutes a form of commodification or a form of ideological recognition of cultural diversity is still an open question. As Regina Bendix emphasizes, each heritagization process must be evaluated in its characteristics, as well *carefull research is necessary to understand the gradations of social control that emerge when a cultural product is transformed in a good of morally and economically enhaced valence* (Bendix 2009, p. 264).

In recent years the relationship between grassroots heritage manifestations in the Metro and regulatory authorities has developed into simultaneously restrictive and inclusive approaches.

On the one hand it has greatly increased the level of institutional control, through interventions of surveillance and repression of any unauthorized activity. At the same time have been implemented public programs in order to promote artistic activities, through the granting of licenses and the creation of institutional spaces for these activities.

The London Underground is organizing since 2003 the project "Sounds of the Underground", a program of musicians selection, to which is released a temporary license to perform in the underground spaces.

Similar programs are held in many subways around the world, including Moscow, Paris, Tokyo, Berlin, New York, Los Angeles, Madrid, Beijing, Barcelona, Toronto and many other cities.

The progressive assimilation of Subculture manifestations in the Metro landscape, both in the form of commodification or ideological appropriation, does not mean that the attrition war with new forms of subversive uses of the normative underground space came to an end. An example is the recent emergence of extreme practices of "Subway parkour", where "Subway surfers" not only perform in acrobatic performances inside the underground space, but perform some very risky maneuvers, such as jumping from one platform to another platform while trains are crossing (Loh-Hagan 2016, p. 23).

This practice, born in the United States, has spread very quickly in Europe and Asia. The recent death of a young "trainsurfer" in the Paris Metro has raised the level of attention of transportation authorities about this phenomenon, until now largely ignored.

3 PROBLEMATIZING THE INTERACTION

Cultural heritage and rapid transit infrastructures interact with each other in many different ways. The interaction can even take many different forms, as well as heritage in Metro spaces can perform many different functions, as a resource of cultural, political and social expression.

Laurajane Smith, stressing the functional value of cultural resources underlines the fact that they are in fact acting as tools or props, used to facilitate the negotiation process of *new ways of being and expressing identity* (Smith 2006, p. 4).

Adopting the idea that heritage is somehow a resource and a tool used for its intrinsic or extrinsic value in the course of socio-cultural negotiation processes, invites us to focus on a different formulation of our opening question: what type of processes, what values and what negotiations are activated inside the Metro by the presence of cultural heritage resources?

We will explore the question by problematizing some main aspects of the controversial relationship between heritage and Metro spaces.

3.1 *Sense of Place and Urban Identity*

Urban identity is a controversial subject that lends itself to a multiplicity of extremely varied theoretical approaches. However, a common element of these approaches is the recognition that in defining the identity of a place, the place cannot be separated from the people who invest meanings in them, while *places are also interpreted, narrated, perceived, felt, understood and imagined* (Norsidah, Khalilah 2015, p. 710).

Nevertheless, in the approach to the issue, there are two different trends placing the focus of the discussion on the material relevance of the sites as witnesses of identity or on the symbolic value of places, as media in the definition process of social identity.

The first case is generally given by approaches aiming to urban governance and spatial planning, as in the case of urbanistic interventions or preservation plans. This sets the focus on qualitative and quantitative characteristics of the relationship between social groups and places.

It is concerned with strategy-making which *seeks to 'summon up' an idea of a city or urban region, in order to do political work in mobilising resources and concepts of place identity* (Healy 2007, p. 23).

Aiming at planning the urban space, this approach implies the possibility that change processes, constantly going on in the urban identity, can be managed by heritage management activities. The methods used in this theoretical approach are meant to identify the socio-spatial components, as well to assess their relevance in the process of change, in order to *designing and control the change* (Kiera 2011, p. 3).

A second, different approach to the theme of urban identity comes from the field of socio-anthropological research, and refers to the concept of "place identity" or "place attachment".

Here the attention is focused more on the symbolic value of places, rather than on their intrinsic value in defining social identity. Place features serve as symbols or icons that can contribute to place identity, and thereby contribute to self-identity of communities. Place attachment refers to the *development of an affective bond or link between people or individuals and specific places expressed through the interplay of affects and emotions, knowledge and beliefs, behaviors and actions* (Norsidah 2015, p. 712).

In other words, it is the social performance creating identity links between landscape markers and community. The urban landscape is, according to this perspective, formed by a mosaic of local landscapes, populated with places to which local people has given symbolic meaning. These meaning are inscribed invisibly into places, and while urban landmarks can be perceived by a social group as typical heritage, *the same landmarks may be an object of indifference or hostility to another* (Tunbridge 2008, p. 236).

Following this perspective, cultural heritage elements play a symbolic role as resources that individuals and communities *mobilise from the archive of memory in the process of forming and expressing identity* (Byrne 2008, p. 169).

3.2 *Mobility: inclusion and exclusion*

Over the past decade a new approach to the study of mobility has been emerging across the social sciences, defined as "New Mobility Paradigm". This research approach set focus on the combined movements of people, objects and information, in all their complex relational dynamics, *and the representations, ideologies and meanings attached to both movement and stillness* (Cresswell 2010, p. 14).

Contemporary human geography studies focus on the history of mobility, its modes of regulation and the power relations associated with it, in short the politics of mobility. Mobility and control over mobility both reflect and reinforce power, while *mobility is a resource to which not everyone has an equal relationship* (Skeggs 2004, p. 49).

Some studies have pointed to the fact that access to mobility constitutes today a form of social divides, in the form of an availability of "mobility capital" (Kauffman, Bergman, Joye 2004), intended as capacity of access networks.

The concept of "network capital" is used to define a *combination of capacities to be mobile, have access to communication devices and secure meeting places; access to vehicles and infrastructures; and time and other resources for coordination* (Elliott, Urry 2010).

The Metro networking factor is an important element in the analysis of social and cultural processes that are triggered by these infrastructures. Processes concerning changes in the perception of urban space by the underground network users, feelings of ownership of this space, the rights of citizenship and ideal residence within these spaces, the relationship between behavioral rules and rights of expression.

In the complex process defining which identities can be included or must be excluded by the mobility network, the demarcation line between public and private space is blurring. The role played by heritage in representing or excluding collective and individual identities, to promote inclusive or exclusive concepts of public space through memory-matter engagements and narratives is one of the main problematic topic regarding the presence of cultural resources inside infrastructural urban mobility projects (Tolia-Kelly, Gillian 2012).

In the case of Berlin Metro, Merrill emphasizes the role of mediation entrusted by the authorities to artistic expressions, in order to re-brand the Berlin U-Bahn as a metaforic reunification factor of the city, after 1989. At the same time outlines how local counter-narrative took place in the Metro, as anti-discourses against the broader background of ideological, political, social and economic change (Merrill 2015).

Merrill marks a boundary in the function of cultural heritage in the Metro: heritage may participate in the Underground negotiation process both as a means of policies aiming at social exclusion or inclusion, as well as a form of dissonant counter-narrative.

3.3 *Image-led planning*

The construction of expensive Rail Rapid Transit systems is generally promoted by public authorities with the main aim to improve urban environmental conditions and promote economic development. However, Higgins and Kanaroglu have recently highlighted the fact that in Rapid Transit projects there is also the search for achieving "intangible planning goals".

The authors underline the fact that *rail transit is often associated with messages of modernity, economic growth and development, global competitiveness, and the attainment of "world city" status as well as a method of giving a city a distinctive identity or brand* (Higgins and Kanaroglu 2016, p. 452).

Planning and building rapid transit systems for improving city image has, according the authors, internal and external purposes.

External purposes are city branding and city marketing ones, important elements in the global competition between city for investments. Internal purposes are the ones linked to the creation of city symbols or myths, aimed to secure public and political support to large governmental initiatives by reducing the tangible complexity of large scale projects into more intangible and easily understood symbolic meanings.

Modern city marketing and branding are phenomena that gained considerable attention in the 1980s. It was mainly concerned with purely promotional activities designed to "sell the city" in an increasing competition between places for *investment, resources, employment, tourism, and human capital in a rapidly globalizing economy* (Higgins and Kanaroglou 2016, p. 454).

Today cities must increasingly sell themselves on their claims to distinctiveness, which means not only the conservation of their mainstream heritage but also the accentuation of their various ethnic spaces (Tunbridge 2008, p. 241). A task largely entrusted to cultural resources.

4 CONCLUSIONS

In this brief analysis we highlighted how complex is the relationship between heritage and Metro infrastructures, as well we set focus on some conceptual aspects of this relationship, challenging the idea of a casual and accessory role of cultural resources in mobility spaces.

On the contrary, we aimed to underline the functionality of heritage resources in giving shape and expressive material to the intangible, political and social objectives that underlie and precede and promote urban mobility projects.

Metro systems have a strong impact on the social fabric of the cities, radically acting on the urban sense of place by creating new connections, new points of orientation and a different physical perception of urban space, creating new connections, new "insides" and new "outsides", a different perception of distances, new points of orientation and a different physical perception of urban space.

Urban mobility infrastructures are offering to urban heritage new expression spaces, but they need also to mediate their highly transformative presence, within the complex and stratified network of "social meanings" that makes up the urban sense of place.

As in all the modern manifestations of the "mobility" phenomenon, the process of negotiating the social meaning of new public transport infrastructures involves inclusive results, causing forms of attraction and consensus, but also exclusive consequences, causing displacement and dissent.

Within these processes, cultural heritage may be playing an extremely important role.

Cultural heritage as symbolic reference can be entrusted with the task of communicating political and social agendas, as well as can be entrusted with a mediation role, by promoting, authorizing and facilitating the presence of the transportation infrastructure in the pre-existing social and cultural urban context.

A better understanding of the functional role of cultural resources in urban mobility projects can contribute to a more reflective and socially oriented approach to the practices of heritage management inside Metro infrastructures.

REFERENCES

Angelo H. and Hentschel C. (2015), Interactions with infrastructure as windows in social worlds: a method for critical urban studies. Introduction, in *City: analysis of urban trends, culture, theory, policy, action*, n. 19, pp. 306-312.

Bendix R. (2009), Heritage between economy and politics. An assessment from the perspective of cultural anthropology, in Smith L. and Akegawa N. (eds), *Intangible Heritage*, Routledge, London and New York.

Bloodworth S. (2014), *New York's Underground Art Museum: MTA Arts and Design*, Monacelli Press, New York.

Byrne D. (2008), Heritage as social Action, in Fairclough G., R. Harrison, J.H. Jameson Jr & J. Schofields (eds) *The Heritage Reader*, Routledge, London & New York, pp. 149-173.

Cervero, R. (1998), The Transit Metropolis. A Global Enquiry, Island Press, Washington, DC.

Chalfant H., Jenkins S., (2014), Training Days: The Subway Artists Then and Now, Thames and Hudson, New York.

Cresswell T. (2006), On the Move: Mobility in the Modern Western World, Routledge, London and New York.

Cresswell T. (2010), Towards a politics of mobility, *Environment and planning D: Society and space*, 28 (1), pp. 17-31.

De Certau M. (1980), The practice of everyday life, eng. Trans. By S. Rendall (1984), University of California Press, Berkeley and Los Angeles.

Davison G. (2008), Heritage: from Patrimony to Pastiche, in Fairclough G., R. Harrison, J.H. Jameson Jr & J. Schofields (eds) *The Heritage Reader*, London & New York, Routledge, pp. 31-41.

Elliott A. and Urry J. (2010), Mobile Lives, Routledge, London and New York.

Feuer W. (1989), Public Art from a Public Sector Perspective, in Art in the Public Interest, edited by Arlene Raven, Ann Arbor, USA, pp. 139-153.

Foucault M. (1975), Discipline and Punish, eng. trans. By A. Sheridan (1977), Pantheon, New York.

Healy P. (2007), Urban Complexity and Spatial Strategies. Toward a relational planning for our times, Routledge, London and New York.

Hebdige D. (1979), Subculture: The Meaning of Style, Routledge, London and New York.

Higgins C. D., Kanaroglou P. (2016), Infrastructure or Attraction? Image-led Planning and the Intangible Objectives of Rapid Transit Projects, *Journal of Planning Literature*, Vol. 31(4),pp. 452-462.

Hill M. (2015), Underground Cathedrals: Moscow's Struggle for a Subterranean Masterpiece, OU, Stalin and Stalinism, history.ou.edu.
Holtorf C. (2005), From Stonehenge to Las Vegas – Archaeology as popular culture, AltaMira Press, Walnut Creek, USA.
Höhne S. (2015), The birth of the urban passenger: infrastructural subjectivity and the opening of the New York City subway, in *City: analysis of urban trends, culture, theory, policy, action*, Volume 19. Issue 2–3, pp. 313-321.
Kauffman V., Bergman M. and Joye D. (2004), Motility: Mobility as Capital, *International Journal of Urban and Regional Research*, Volume 28.4, pp. 745–756.
Kiera A. (2011), The local identity and design code as tool of urban conservation, a core component of sustainable urban development – the case of Fremantle, Western Australia, *City & Time*, 5 (1), 2.
Lefebvre H. (1991), The Production of Space, English translation by Donald Nicholson-Smith, Blackwell Ed., Oxford.
Loh-Hagan V. (2016), Extreme Parkour. Nailed it!, Cherry Lake Publishing, Ann Arbor, Michigan, USA.
Lowenthal D. (1998), The Heritage Crusade and the Spoils of History, Cambridge University Press, Cambridge.
Merrill S. (2015), Identities in transit: the (re)connections and (re)brandings of Berlin's municipal railway infrastructure after 1989, in *Journal of Historical Geography*, Volume 50, pp. 76–91.
Nadolny A. (2015) Henri Lefebvre's concept of urban space in the context of preferences of the creative class in a modern city, in *Quaestiones Geographicae* 34(2), Poznań, pp. 29–34.
Norsidah U., Khalilah Z. (2015), The notion of Place, Place Meaning and Identity in Urban Regeneration, in *Environmental Settings in the Era of Urban Regeneration*, Procedia 170, Elsevier, Amsterdam, pp. 709-717.
Parlama L., Stampolidis N. (2001), Athens: The City Beneath the City. Antiquities from the Metropolitan Railway Excavations, Harry N. Abrams, New York.
Sheller M. and Urry J. (2006), The New Mobilities Paradigm, *Environment and Planning*, volume 38, pp. 207-226.
Skeggs B. (2004), Class, Self, Culture, Routledge, London and New York.
Smith L. (2006), The Uses of Heritage, Routledge, London and New York.
Smith L. (2008), Towards a Theoretical Framework for Archaeological Heritage Management, in Fairclough G., R. Harrison, J.H. Jameson Jr & J. Schofields (eds) *The Heritage Reader*, London & New York, Routledge, pp. 62-74.
Tunbridge J.E. (2008), Whose Heritage to Conserve? Cross-cultural reflections on political dominance and urban heritage conservation, in Fairclough G., R. Harrison, J.H. Jameson Jr & J. Schofields (eds) *The Heritage Reader*, London & New York, Routledge, pp. 235-255.
Whyte W. H. (1979), The Social Life of Small Urban Spaces – The Street Corner, Conservation Foundation, Washington, DC.
Wolmar C. (2012), The subterranean railway: how the London Underground was built and how it changed the city forever, Atlantic Books, London.

Tunnels and Underground Cities: Engineering and Innovation meet Archaeology, Architecture and Art, Volume 1: Archeology, Architecture and Art in underground construction – Peila, Viggiani & Celestino (Eds)
© 2020 Taylor & Francis Group, London, ISBN 978-0-367-46574-2

The archeological evidences of the De Amicis Station in the Milan Metro line 4, Italy

G. Lunardi, G. Cassani, M. Gatti & S. Gazzola
Rocksoil S.p.A., Milan, Italy

ABSTRACT: Line M4 will cross Milan with a length of about 15 km from west to east along Viale Lorenteggio, through the south of the old town and along the axes of Forlanini up to Linate Airport. The central part is very close to the historical town and the interference with the pre-existing, especially the monumental and archaeological heritage, shall be carefully investigated. In the stretch between Tricolore and Solari Stations, the metro stations are considered as "deep", reaching depths of 25-30 m below ground level: the stations layout is composed by a central shaft, with transverse dimensions about 10 m located in the main roads or parks, and lateral running tunnels, constructed using a TBM-EPB with a 9.15 m diameter which is able to be lodged inside the tunnel the stations' platforms. The "De Amicis" station is one of these; during the first excavation step of De Amicis Station, an ancient wall which constituted the Naviglio of San Gerolamo, dated back to the Middle Ages (XIII-XIV sec.) have been discovered. Some stones of the wall are from the Roman Era, probably part of a previous structure, reused as a "quarry". The relevance of these findings, located close to the Pusterla area, one of the secondary entrances of the city, forced to modify the layout of the station to create a museum space to host the ancient wall in its original position. The paper describes the "deep station" project, with construction details, and the solution adopted to exhibit the archaeological evidences.

1 INTRODUCTION

The paper deals with the construction of the De Amicis Station of the M4 Underground Project in Milan. After a general description of the entire M4 project, it will be presented the original De Amicis project, designed considering existing buildings and roads before any type of excavation, and the variation project, defined after first excavations and the finding of an old wall, related to the ancient Naviglio of San Gerolamo dated back to the Middle Age (XIII-XIV sec.).

The variation project has been defined in order to valorize the particular finding; for this reason, almost the entire layout of the station has been modified. The paper will also provide information about the management of the finding regarding especially the removal of the ancient wall. The construction works are performing by the "MetroBlu" consortium, composed by Astaldi and Salini-Impregilo.

2 METRO LINE M4

The construction of the new M4 (blue) line, managed according to the project finance formula, has been divided into three different sections. The first section will link Milan Linate airport to the Forlanini railway station and then to the Tricolore Station. The second section will connect the Tricolore Station to Parco Solari Station. The third section will connect San Cristoforo railway stations to section two. The new metro line will link the eastern and western parts of the city and its construction will be completed by 2022. The line has altogether

21 underground stations and will have a total length of 14.2 km. The running tunnels will be bored using six EPB TBMs: 4 TBM for the two single-track tunnels with diameters of 6.5 m for the sections outside the city centre (2 from east and 2 from west). The central part of the track will be constructed using two 9.15 m diameter EPB TBMs, to include the platform for the subway stations.

2.1 *The project*

Line M4 will cross Milan with a length of about 15 km from west to east along Viale Lorenteggio, passing south of the old town and along the axes of Indipendenza, Argonne and Forlanini up to Linate Airport (Figure 1). The M4 route will optimize city coverage, loading options, and interconnection with the metro and suburban rail network, thereby improving the overall network effect of the entire public transport system in the city.

Line M4 will be a "fully automated light rail" system, driverless, and with automatic platform doors and a CBTC (Communication Based Train Control) signalling system. The trains will be 50 m long, considerably shorter than rolling stock in circulation today. Likewise, the 50 m long stations will also be shorter than the 110 m stations on lines M1, M2 and M3. The relatively compact dimensioning of the structures, particularly the stations, means that construction work on the line can be carried out more easily and with less impact. The automation of the system will ensure higher frequencies for the vehicles (90 seconds, theoretically reducible down to 75) providing the capacity to transport 24,000 to 28,000 passengers per hour per direction. The new line M4 will pass through neighbourhoods with high population densities, so the construction methods have been planned to minimize impact at the surface and adapt to an underground affected by a great amount of infrastructure and by the presence of a significant amount of water. The extensive use of mechanized tunnelling, and the selection of a single-track twin tunnel layout help maximize the flexibility and adaptability of the route, which is situated entirely underground except for the depot/office area. There are currently two interchanges with existing Metro lines, one with the red line at San Babila station, and one with the green line at S. Ambrogio station. In the future, there will be three interchanges with suburban railway lines, one with Lines S5, S6 and S9 at Forlanini FS station, one with Lines S1, S2, S5, S6, S13 at Dateo station, and one with Line S9 at San Cristoforo station, where there is also a connection to the Milan-Mortara railway. Lastly, an interchange with Linate airport is planned.

Most of the underground construction on the route will be carried out by mechanized tunnelling with the use of two TBM geometries, one with a bored diameter of 9.15 m and the other with a bored diameter of approximately 6.36 m. The TBMs with diameters of

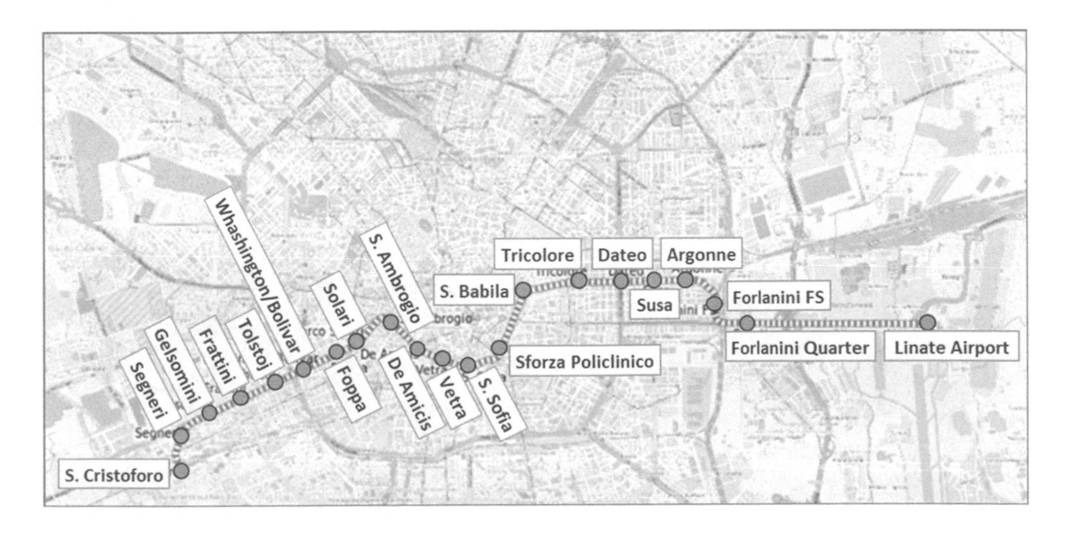

Figure 1. Metro Line 4 Layout.

approximately 6.36 m will be used for the sections from Manufatto Ronchetto, in the San Cristoforo area, to the Parco Solari station and from Linate Airport to the Tricolore station. The TBM with a diameter of 9.15 m will be used for the section from the Parco Solari to the Tricolore station. As indicated, the machine with a diameter of 9.15 m will be used in the section through the deep stations in the historic centre in order to enable the installation of the station platforms directly inside the inner contour of the tunnel in segments. This allows a considerable reduction of the impact on existing structures compared to the use of conventional methods of tunnel excavation, after performing consolidation work (Lunardi 2017).

2.2 *Running Tunnels*

The running tunnels for Line M4 of the Milan Metro will all be constructed using EPB (Earth Pressure Balance) TBMs. The final lining of the tunnel will be made of precast segments placed by the machine immediately after excavation at a small distance behind the face. The ring for the 6.36 m diameter tunnels is composed of six segments (5 + 1 keystone) with a thickness of 28 cm. That of the 9.15 m tunnel is composed of seven segments (6 + 1 keystone) with a thickness of 35 cm (Figure 2.a).

The final lining, in addition to performing and ensuring the normal function of support in both the short and long term, must also provide the required hydraulic seal. For this reason, the segments are fitted with watertight neoprene seals along all surfaces in contact with other segments, and arranged in corresponding housings on the sides of the segment, to ensure the required water-tightness under hydrostatic pressures with the planned clamping forces. In order to ensure water-tightness between the segments of adjacent rings, as well as for reasons of safety during the transitory phases of handling and laying the segments themselves, the connection is provided by means of longitudinal mechanical dowels (Biblock System or equivalent type) arranged at regular intervals around the circumference.

The 6.36 m TBM is equipped with 16 thrust plates arranged in groups of three for each segment plus one for the keystone, or a total of 32 jacks acting in pairs on each plate. The dimensions of the plate are 26 × 70 cm. The maximum thrust that the machine can exert is equal to 42,575 kN or 2,660 kN per thrust group. The 9.15 m TBM is equipped with 19 thrust plates arranged in groups of three for each segment plus one for the keystone, or a total of 38 jacks acting in pairs on each plate. The dimensions of the plate are 33 × 100 cm. The maximum thrust force that the machine can exert is equal to 81,895 kN or 4,310 kN per thrust group (Lunardi, 2017).

2.3 *Stations*

For the construction of stations and structures, the planners tried to resort as much as possible to the open excavation method supported by reinforced concrete diaphragm walls (open bottom-up method), which is compatible with the existing road network and construction site areas required for the execution of the works. This type of construction is applicable in the Linate–San Babila axis since the bodies of the stations are manufactured within the central

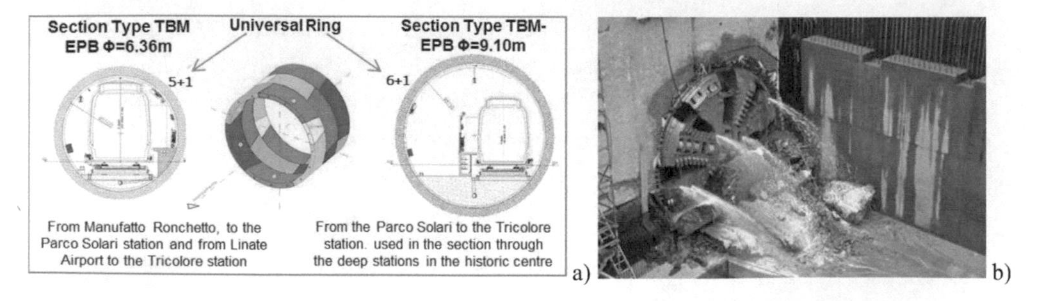

Figure 2. a) TBM Sections type; b) Forlanini Railway Station: TBM breakthrough odd track.

parterre of the boulevards, minimizing interference with road traffic while providing sufficient construction site areas. The same type of construction will also be used for the stations on the San Cristoforo–Parco Solari section, with the exception of Gelsomini and Segneri stations, for which totally or partially closed bottom-up methodology is planned.

The Line 4 stations, with the exception of those in the central section, generally have a limited depth of about 15 m, through which the two TBMs pass "unloaded". The level of the floor slab thus remains occupied by the TBM supply site comprising the rails for trains transporting precast segments, conveyor for transporting the muck out of the tunnel, hoses for cooling water and a medium voltage cable for the TBM power supply, for each of the two machines (Figures 3a, b). The use of semi-precast self-supporting structures is planned constructing the horizontal elements of the station. The supporting structures for the excavations, bulkheads, tie rods, and bottom sealing blocks, will be installed by means of clamshell excavation and stabilized with bentonite slurry, tied with anchors outside the aquifer and bored without the use of a preventer and bottom buffers in integrated cement injections.

The stations considered as "deep" are Sant'Ambrogio (Figure 4a), De Amicis, Vetra, Santa Sofia, Sforza Policlinico and San Babila. The functional installation is composed of a central

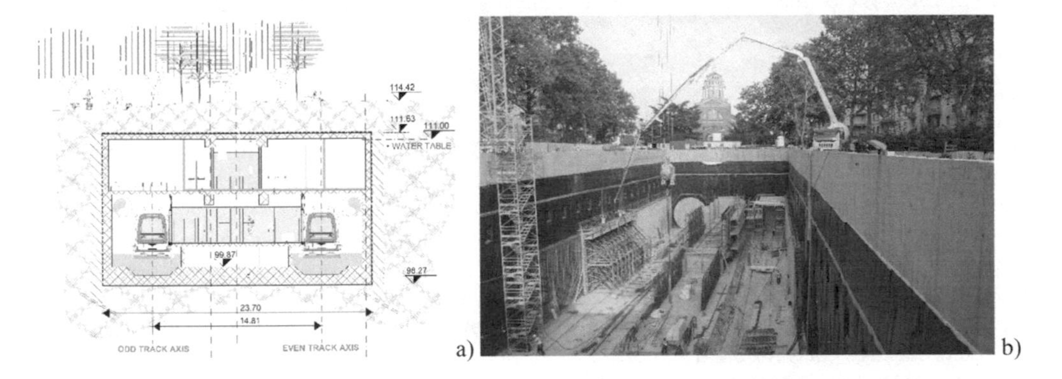

Figure 3. Argonne Station: a) Detail of the reinforced concrete piling; b) general view.

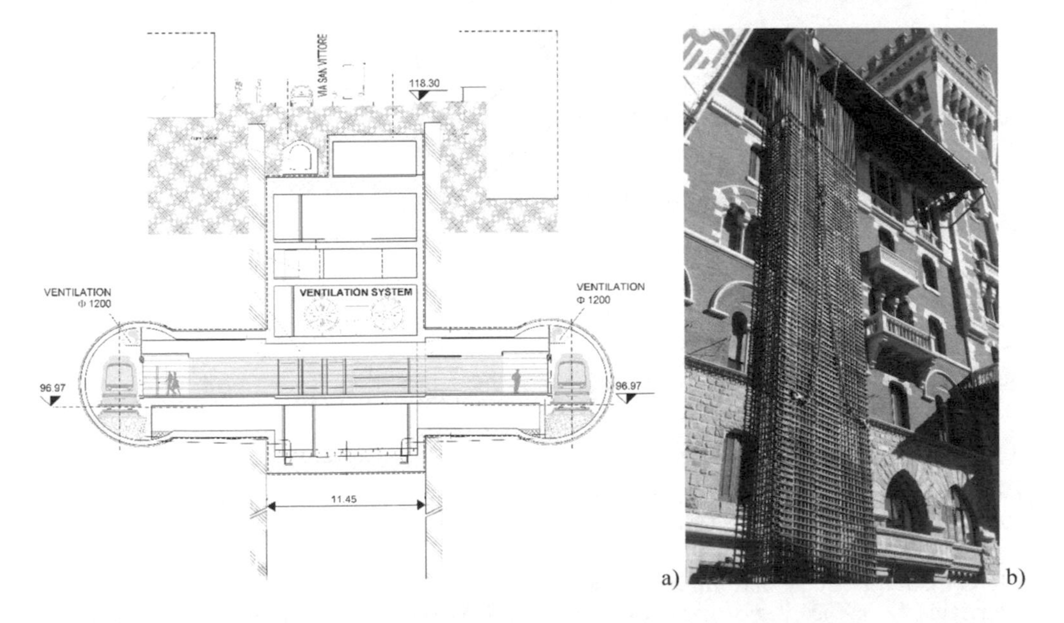

Figure 4. a) Sant'Ambrogio deep station - interconnection with Line 2; b) detail of the reinforced cage.

shaft with an excavation depth up to approximately 30 m and transverse dimensions limited to approximately 10 m. Outside this, running tunnels are constructed using a TBM with 9.15 m diameter, which is sufficient to accommodate the platforms of the stations. The station have been excavated by means of lateral retaining walls, represented by RC diaphragm, supported by anchor or steel frame (locally by slabs, for top-down system) and with a grouted plug at the base able to prevent water inflow. This plug for the deep stations are pushed to a consistent depth below the bottom of the excavation, up to 16–17 m, bringing the total length of the retaining walls up to approximately 50 m, with a lot of construction problems (connections for reinforced cages, amount of reinforcement, Figure 4b). The connecting bypass between the station shaft and the platforms will be excavated after the soil has been grouted from ground level with the continual use of cement and silicate mixes (Lunardi 2017).

The project included an analysis of the interference between the excavations of the running tunnels with the surface buildings and structures. The subsidence basin were evaluated in detail relatively to the excavation. Furthermore, the subsidence/displacements were determined at ground level as well as at the level of the foundation, assessing the damage class expected for each building. In subsequent stages of the project, on the one hand, the interference analysis will be more highly developed with numerical analysis methods, and on the other hand, the detailed findings for some of the buildings in the area of the subsidence trough will be completed.

2.4 *The interferences*

The hazards associated with the tunnel construction in urban areas include poor ground conditions, presence of water table above the tunnel, shallow overburden and ground settlements induced by tunneling with potential damage to the existing structures and utilities above the tunnel. A fundamental aspect of metro line design is the evaluation of the excavation interference (Cassani, 2005; Mancinelli, 2009). In case of M4 metro line a study to assess the interference between underground excavations and the existing buildings has been developed. The excavation takes place with a mechanized system, adopting an EPB type TBM. The subsidence basins and the settlements related to the excavations have been evaluated to assess the expected damage class for each building.

The project included an analysis of the interference between the excavations of the running tunnels with the surface buildings and structures. The subsidence basin were evaluated in detail relatively to the excavation. Furthermore, the subsidence/displacements were determined at ground level as well as at the level of the foundation, assessing the damage class expected for each building. In subsequent stages of the project, on the one hand, the interference analysis will be more highly developed with numerical analysis methods, and on the other hand, the detailed findings for some of the buildings in the area of the subsidence trough will be completed.

3 DE AMICIS STATION & THE AERCHAEOLOGICAL FINDINGS

One the deep station is located at De Amicis Street. During the first phases of construction of this station, realizing the guidance structures for diaphragms excavations, an ancient wall dated back to the Middle Age (XIII-XIV sec.) has been found. These findings forced to revise the station layout as described in the following. Preliminary to the intervention description, a brief historical review and a description of the archeological findings are proposed.

3.1 *The archaeological finding of the M4 Line*

The central section of the M4 line between S. Babila and S. Ambrogio stations touches locations of major importance for the historical memory of Milan, following the periphery of the Roman city of Mediolanum and meeting the city walls, the major roads out of the town, the early Christian basilicas and the cemeteries.

Particularly in the sites of these two stations, S. Babila and S. Ambrogio, touching testimony of the identity and history of the inhabitants of two millennia ago came to light: tombs of men, women and children containing objects of daily life, preserved intact a few feet below today's streets and pavements.

Around the stations of S. Sofia, S. Calimero and Vetra, traces were discovered of the complex network of canals and bridges. Near of the De Amicis station, close to Piazza Resistenza Partigiana, mediaeval defense works were recovered together with a bridge over the historical S. Girolamo canal.

The most touching testimonies that reveal the identities and stories of two thousand years ago emerged in the construction sites of the San Babila and Sant'Ambrogio stations: tombs of men, women and children, sometimes buried with a trousseau that has been preserved intact at a shallow depth, under the sidewalks and roads that people tread daily. While the work on the construction site of the S. Ambrogio station were in progress, in the sites for the maintenance of the "TBM", located in the churchyard of San Vittore al Corpo and the Basilica of S. Ambrogio archaeological excavations unearthed some burials (Figure 5), expected on the basis of the archaeological surveys. The churchyard of San Vittore al Corpo was the location of an Imperial Mausoleum dates back to IV sec. and usually around the Mausoleum were buried the important persons of that time.

Another important archeological item is the Milan historical water system, which was developed since X-XI century b.C. during Middle Ages period. It had principally defense function and it was also used for commercial reasons. The water system later developed and increased its extension but principally it kept the same geometry until the coverage of XX century, after 1929 (Lunardi, 2018).

The first layout of Navigli also included the zone of the actual Carducci and De Amicis streets. On De Amicis Street, the ancient Roman brick wall has been discovered and it seems that it was part of the ancient water course of the Medieval Naviglio, the Inner Circle of Navi-gli ("Cerchia interna dei Navigli"). Figure 6a represents the Milan's Medieval Map and of its water courses: Figure 6b shows the actual position of the De Amicis Station and it can be seen that it's positioned just over the ancient Naviglio of San Gerolamo, from Carducci to De Amicis Street and continuing on Santa Sofia Street.

Photos taken at the beginning of the 20th century shows the road axe Carducci, De Amicis and Molino delle Armi streets with water courses still open could help knowing how and where the old Naviglio was placed (Figure 7a). Figure 7b is a street view of the actual situation.

Figure 5. San Vittore al Corpo churchyard: a) archaeological excavations; b) archeologist compiling the anthropological data sheet; c) unearthed burials.

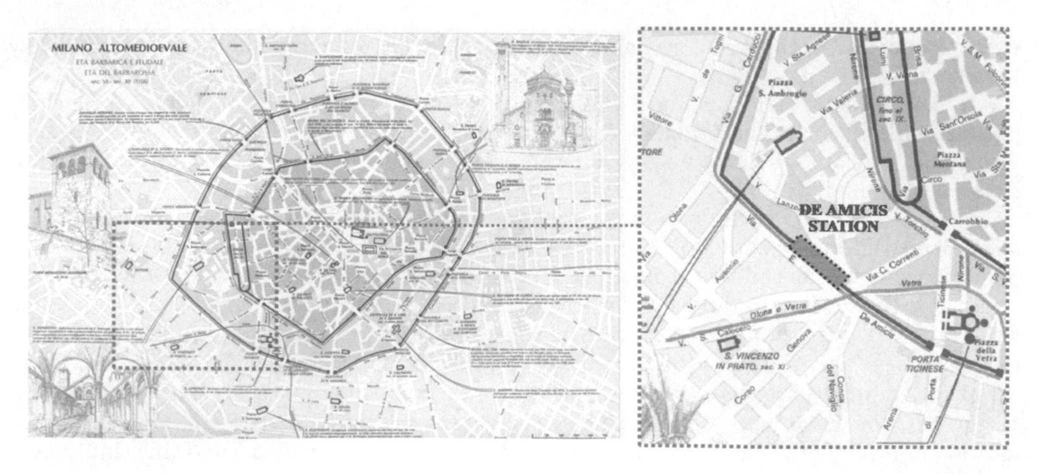

Figure 6. a) Medieval map of Milan and of its water courses X-XII century b) detail view with the overlap of the De Amicis Station.

Figure 7. Road axe Carducci, De Amicis and Molino delle Armi a) at the beginning of the 20th century b) street view of the actual situation.

3.2 *The original station design*

De Amicis Station is located in the central part of the M4 track along via De Amicis close to the crossing between via Correnti and Corso Genova (Figure 8a). The station is approximately 70 m long and 12 m large for a medium depth of 30 m. The tunnels are approximately 25 m deep.

The station is realized between concrete diaphragms wall 40-42 m long, 1 m thick. The construction is realized initially, after few meters of excavations, by the cast of the concrete top-slab of the station. Then, 4 orders of steel struts and perimetric beams are provided in order to reach the bottom of the station. Afterwards, the base floor is casted and, going towards the surface, all the other floors, removing each order of steel retaining structure (Figure 8b).

3.3 *The Ancient Roman wall & station design revision*

During the excavation for diaphragms of De Amicis Station, an ancient wall dated back to the Middle Age (XIII-XIV sec.) has been found. The wall belonged to the Naviglio San Gerolamo and an initial analysis of the mortar dates back to a period between the thirteenth and sixteenth centuries (Figure 9).

The importance of its discovery lies in the particular techniques with which it was built and in the position in which it was found, at the height of the Pusterla: one of the secondary accesses to the city. The roman wall was an unexpected discovery and it will be preserved,

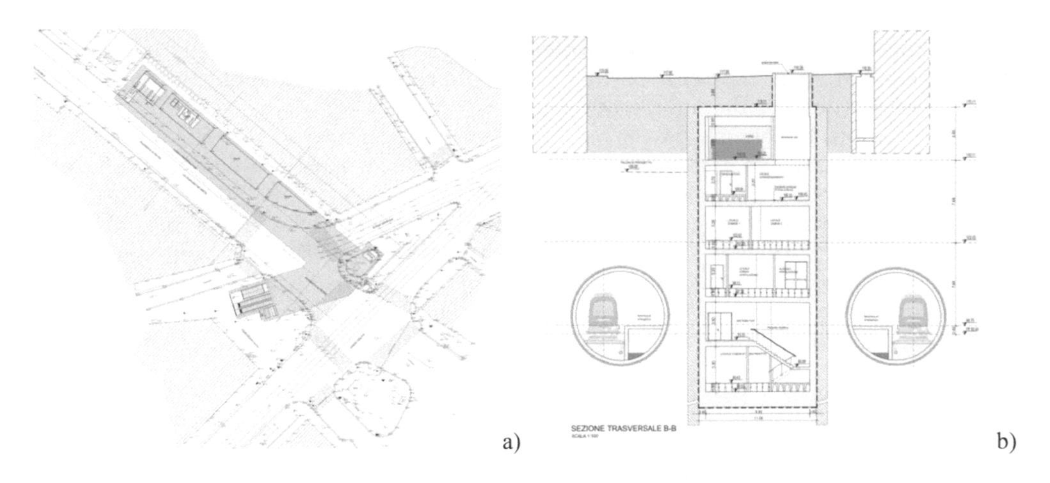

Figure 8. Original project of De Amicis Station: a) layout; b) transversal section.

Figure 9. Roman wall unearthed during the excavation for De Amicis Station.

enhanced and exposed transforming the atrium of De Amicis station into a museum hall. The monument, 12 metres long and 3 metres high, will located immediately before the turnstiles, putting back in its original position inside the M4 station.

The wall has been completely removed from its original position cutting it by diamond wires (Figure 10a), waiting to came back in the same place when the station will be completed.

Figure 10. a) Diamond wire used for wall cutting; b) Roman wall detail.

According to these findings, an archaeological revision of the station layout has been proposed. The intent of the project is hosting the historical wall in the station so that passengers can see it while attending the underground line.

For this reason, the original layout of the station has been revised defining a specific area to store the wall: from an archaeological point of view the original position of the historical find has been maintained the same (Figure 11).

In the revised project, along the De Amicis Street side, there's now a superficial body, just over the odd tunnel (Figure 12). This part of the station is realized between concrete diaphragms and internal structures are provided with a precast coverage in order to foresee different yard phases, especially related to traffic superficial deviations. During station's construction, in the last phase, after some demolitions of existing retaining structures, the

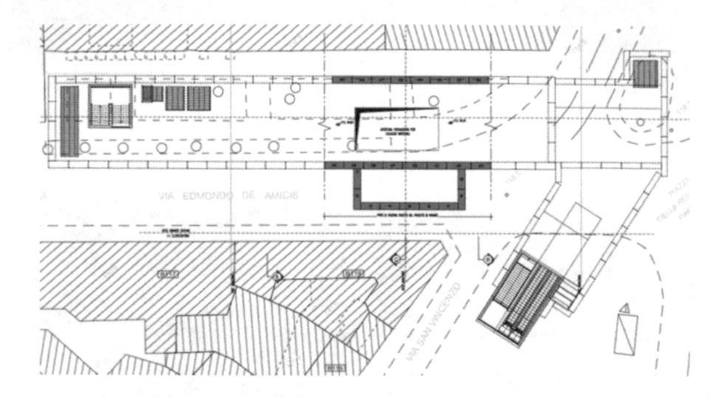

Figure 11. Plan of the revised station project.

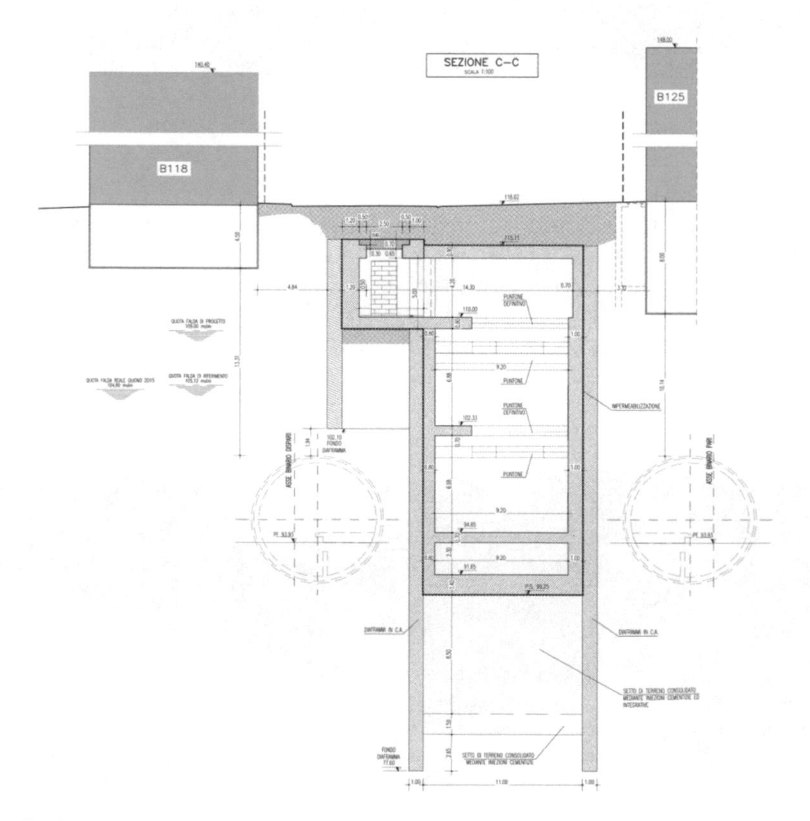

Figure 12. Section of the revised station project, with the position of the ancient wall.

precast coverage will be removed in order to place parts of the historical wall. In the following Figure 12, the revised project is presented.

4 CONCLUSION

The paper presents a general description of the M4 underground project in Milan. Some information about the alignment in its various parts, about the TBMs involved in excavations and about underground stations are also reported.

The central part of the M4 line interests the downtown of the old city of Milan and archeological findings were found during the investigation period and the first time of construction. This especially between S. Babila and S. Ambrogio stations, which touch locations of major importance for the historical memory of Milan, following the periphery of the Roman city of Mediolanum and meeting the city walls, the major roads out of the town, the early Christian basilicas and the cemeteries. Another important historical item is water system, which was used for defense function and for commercial reasons.

A significant archeological finding is located at the site of the De Amicis Station: during the first excavations activities, an ancient wall dated back to the Middle Age (XIII-XIV sec.) has been discovered. It's a part of the ancient Naviglio of San Gerolamo.

According to these archaeological findings, the original station's project was fully revised. The intent of the new project is to host the ancient Roman wall trying to keep it, as possible, in its original position and valorizing it, so that passengers are able to visit it while attending the underground metro. Some details of the wall, during its excavation, and of the new station layout are presented.

The De Amicis station it's a great example: when engineer and archaeologist work together, they are able to build a bridge between past and future, so that the today's subway passengers can see the past's passenger route.

REFERENCES

Cassani G., Mancinelli L., 2005. Monitoring surface subsidence for low overburden TBM tunnel excavation: computational aids for driving tunnels. *IACMAG (International Association for Computer Methods and Advances in Geomechanics) Conference on prediction, analysis and design in geomechanical applications*, Torino.

Lunardi G., Cassani G., Gatti M., Zenti C.L., 2018. The role of underground transportation inside Milano's Smart City perspective. *In*: *Proceedings of ITA-AITES World Tunnel Congress 2018*. Dubai, United Arab Emirates.

Lunardi G., Cassani G., Gatti M., Gazzola S., 2017. Milan line 4: from east to west crossing the downtown. *In*: *Proceedings of AFTES International Congress "The value is underground"*, Paris, France.

Mancinelli L., Gatti M., Cassani G., 2009. Numerical simulation of an excavation near buildings. *In*: *Proceedings of ITA-AITES World Tunnel Congress*, Budapest, Unghery.

http://www.metro4milano.it

https://milano.corriere.it

https://www.ilgiorno.it

Tunnels and Underground Cities: Engineering and Innovation meet Archaeology, Architecture and Art, Volume 1: Archeology, Architecture and Art in underground construction – Peila, Viggiani & Celestino (Eds)
© 2020 Taylor & Francis Group, London, ISBN 978-0-367-46574-2

Archaeology and tunnelling interaction in the railway project of Catania underpass in Sicily, Italy

E. Manfredi, F. Iannotta, F. Romano & S. Vanfiori
Italferr S.p.A., Roma, Italy

ABSTRACT: This paper deals with the preliminary design of the underground railway from Catania main station (Catania Centrale) to Catania Acquicella station, as part of the modernization of the Messina-Catania-Palermo line in Sicily (Italy). The new project involves the upgrading of the existing link through the construction of a new double-track EPB-TBM bore tunnel 1.3 km long with a diameter of 10 m. This will restore to the city the original urban layout, deeply modified by the existing old surface line built at the end of the 18th century. The feasibility studies conducted over the last few years have revealed interference with archaeological finds and the project has had to be adjusted due to the presence of the buried remains. Therefore, specific historical studies and geo-archeological surveys were carried out to interpret the stratigraphic data and to reconstruct the soil layers possibly containing archeological findings.

1 INTRODUCTION

The doubling of the Catania railway line between Catania Centrale station and Catania Acquicella station is about 5.6 km long. The alignment runs almost entirely in underground (cut&cover and bored tunnels) below the city of Catania and entails the construction of the new underground Catania station, of two intermediate stations (Porto/Duomo and S. Cristoforo) and of the receiving Fontanarossa station.

The project wholly involves the urban area of Catania, with the aim of improving urban public services and rehabilitating the downtown area currently occupied by the railway.

The preliminary design is currently under way and envisages the dismission of the historical line between the Catania Acquicella station and the harbor, in accordance with the requalification project of the Waterfront planned by the Municipality of Catania.

The design is highly complex because the underground works involve a highly urbanized area in a heterogeneous and very complicated geological and hydrological context. Moreover the territory has been marked by intensely inhabitation since the ancient Greeks until today and by catastrophic events, which has led to a complex juxtaposition of buildings, deeply conditioning its current layout. One of these natural events is the impressive eruption of Mount Etna Volcano in 1669 that produced a lava flow that reached the city and even changed the ancient coastline (Figure 1). The layer of lava will be intercepted by the tunneling works for most of the alignment.

The new underground works cross through an area holding significant archaeological remains, the most important of which is the ancient Greek-Roman baths called '*Terme dell'Indirizzo*'.

This paper illustrates the results of the preliminary studies and surveys carried out in order to acquire a comprehensive picture of the geology and archaeology of the area crossed through by the new alignment (Figure 2). The archaeological risk assessment allowed to develop the design in such a way that to reduce to a minimum their impact on the city's important historical heritage.

Figure 1. Historical depiction of the lava flow of 1669.

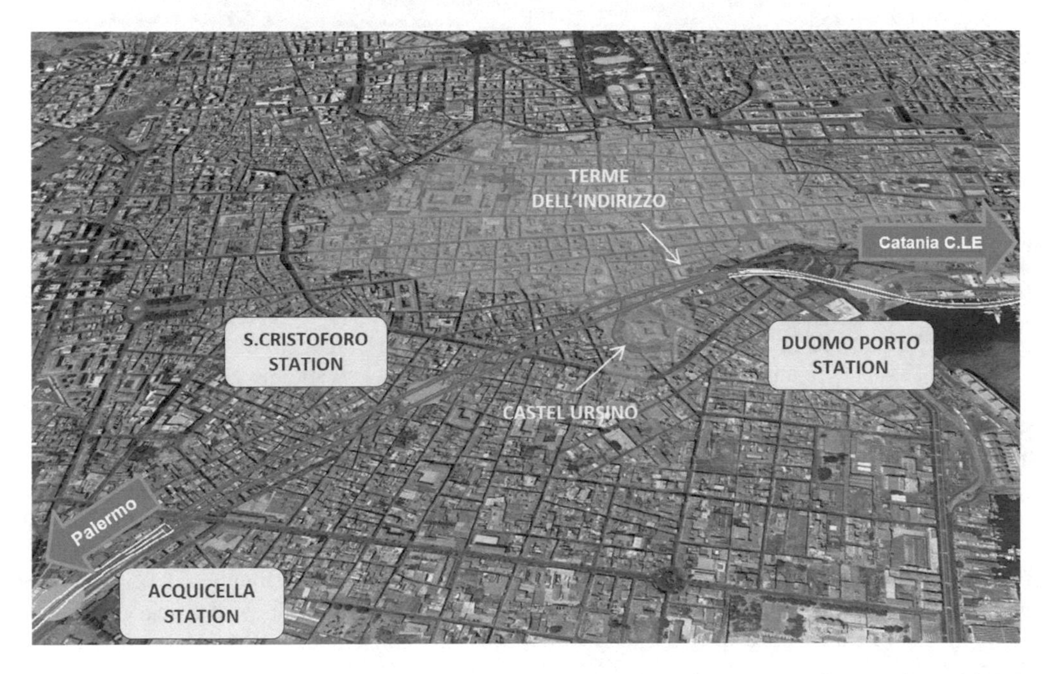

Figure 2. Chorography showing the railway line and the location of the new stations envisaged by the project.

2 GEOLOGICAL AND STRATIGRAPHIC SETTING

The project area is located in the South-West sector of the volcanic system of Mount Etna and it is characterized by a thick regressive Pleistocene succession, on which the volcanic products of the Etna and the Quaternary deposits of the Piana di Catania plain and the Ionic coast are laid in unconformity (ISPRA 2009; Monaco & Tortorici 1999).

The lithostratigraphic layout was reconstructed based on the analysis of all the geological and geotechnical investigations specifically carried out for the project. The substrate belongs to the Lower Pleistocene "Argille grigio-azzurre" Formation made up of grey-blue clays ("FAG"

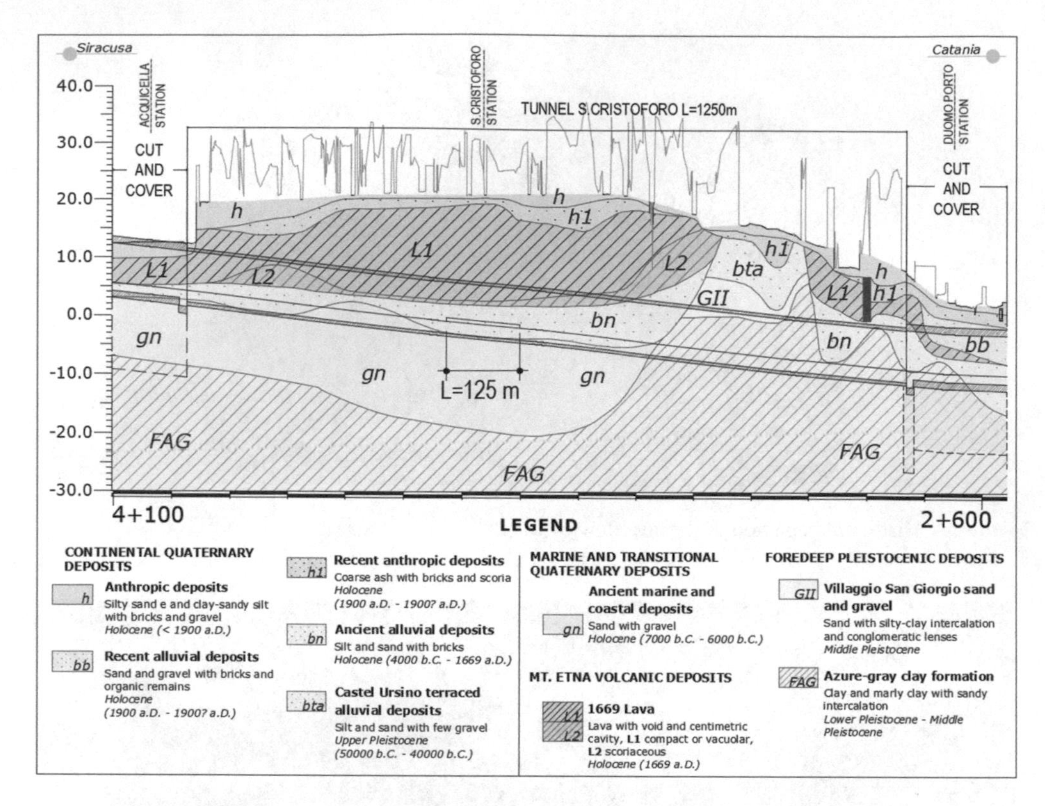

Figure 3. Geological profile of the bored tunnel.

in Figure 3), covered in alternation by the Middle Pleistocene “Villaggio S. Giorgio” Formation (“GII” in Figure 3), consisting of sands interbedded with clay and silt and conglomerate lenses.

In Castel Ursino area, these formations of marine origin form the so-called ‘morphological high’ a relief on which the terraced alluvial deposits of Castel Ursino (“bta” in Figure 3) rest in stratigraphic unconformity. The alluvial are made up of sandy silts and silty sands with volcanic elements and local gravels.

In the southern area, the ancient marine and coastal deposits (“gn” in Figure 3), mostly sands and silty sands, rest on the clayey substrate. The ancient alluvial deposits (“bn” in Figure 3) made up of silty sands and sandy silts, interspersed with numerous fragments of bricks, are present above the marine and coastal deposits. The findings of bricks in alluvial deposits prove that this level is the only one where one can expect to find archeological evidence sealed off by the lava flow of 1669 (“L” in Figure 3) which, after bypassing the ‘morphological high’ of Castel Ursino, reached the ancient coastline.

The strata which potentially include archeological findings are therefore the alluvial deposits (“bn” in Figure 3) laying above the Pliocene strata. From the archeological viewpoint, it was necessary to ascertain, among other things, the extension and the depth of these strata using additional surveys in the *Terme dell’Indirizzo* area.

3 ANALYSIS OF THE HISTORICAL AND ARCHEOLOGICAL CONTEXT

Within the Design phase an Archeological Survey was prepared with the aim of assessing the distribution, the entity and the consistency of the historical and archeological assets, and of identifying the areas where it was possible to foresee interference between the design solutions and any existing or expected archeological deposits.

Figure 4. Map of archeological findings within the city of Catania.

The new railway line posed the serious issue of crossing through the urban area within the 16th century town walls and from the harbor area to the ancient *Terme dell'Indirizzo* baths complex.

The archeological sites intercepted along the way in the central area of the city are many and belong to various historical periods. Almost all of the excavations carried out in the city showed complicated and multi-stratified situations.

This paper do not dwell on the historical and architectural vicissitudes of the city of Catania, or on the architectural evidence still part of the urban texture, but it examines further in depth the results of the investigations performed in order to overcome the critical issues of the interference between the project and the city's precious historical and archeological assets.

The impact with the archeological assets within the 16th century town walls has been partially solved by keeping the tunnel mainly inside the Pleistocene deposits, while the alignment adjacent to the *Terme dell'Indirizzo* complex required further study.

In order to solve the problem of crossing through this portion of the city all the designers involved closely worked together with the Councillorship for the Cultural and Environmental Assets of Catania, in a virtuous process aimed at solving any works impact as the possible interference of the tunnels with the delicate structures of the ancient building complex.

Therefore, four additional archeological drillings were carried out in the area located between the *Terme dell'Indirizzo* baths and the ancient coastline so as to establish with greater precision the geo-archeological profile of that portion of urbanized land and to forecast the possible impacts due to tunneling.

3.1 *Additional coring to assess the impact on the Terme dell'Indirizzo baths complex*

The boreholes, called S1-S2-S3-S4 (Figure 5), were performed in the harbor of the ancient city of Catania. Boreholes S1 and S2 are located in an area that in ancient times was close to the sea, while boreholes S3 and S4 are located in the 'morphological high'.

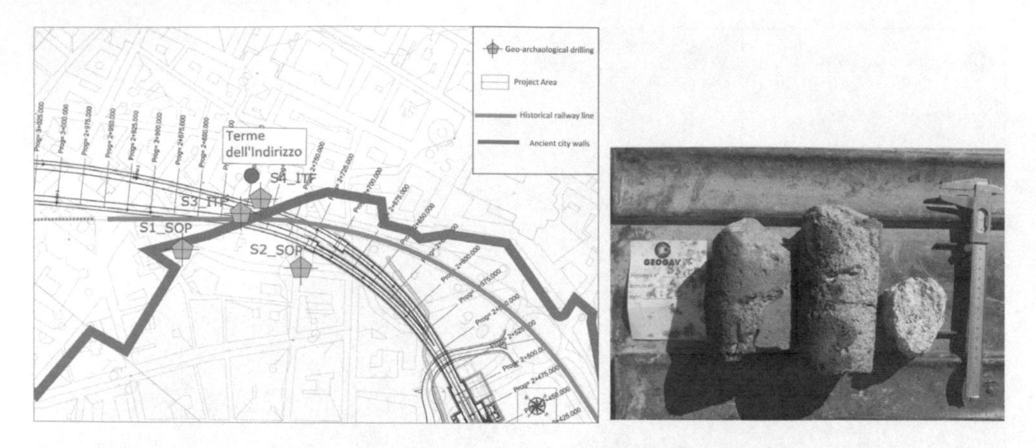

Figure 5. Additional boreholes locations. Fragments of cement wall found in core sample S3.

Without a doubt the most important archeological structure is the complex of buildings called *Terme dell'Indirizzo*, located in Piazza Currò, that is one of the city's most important monuments from the Roman Imperial era.

The boreholes depths at which archeological evidence was found confirmed the hypothesis that that area was close to the sea.

Core sample S3 intercepted a concrete wall that probably belonged to the 16th century town walls. The wall is located between 3.20 m and 11 m below the ground level.

The wall stands on ancient alluvial deposits (bn) that, even though it was not found any evidence of anthropic existence, could contain archeological elements.

It should be noted that over the centuries the district of the *Terme dell'Indirizzo* has undergone various and consistent town-planning changes that have considerably modified its layout and stratigraphy. Recent archeological investigations show that the structures visible in the south area of the baths complex, the ones closest to the new railway line, are not a part of it. Moreover, although in that sector the archeological investigations have reached depths greater than 2 m, there was not found any stratigraphy belonging to the ancient Roman period.

In the early 16th century, the convent of the Carmelites was built close to the complex and in some places on top of it, in an urban context that most probably already included the town wall intercepted by core sample S3.

In ancient Roman times, the baths stood in a higher area (morphological high) close to the sea and accessed from the harbor via an uphill road. Therefore, the new railway tunnels located further below the ancient link do not interfere with the baths complex.

The S4 borehole, drilled outside the town walls, revealed the layer of lava flow dating back to 1669 (L), at a depth between 6.70 m and 8 m below ground level, and then the layer of ancient alluvial deposits (bn) without archeological evidence, about 3m thick.

4 ARCHEOLOGICAL RISK ASSESSMENT

The main purpose of the archeological survey was to assess the 'relative archeological risk' based on the type of structure to construct, on the geological profile, on the entity of archeological findings and on the distance between the latter and the works. The additional data coming from the coring surveys conducted in the *Terme dell'Indirizzo* baths complex area allowed to determine with greater certainty the archeological risk to be considered in designing towards overcoming the critical issues.

4.1 *Analysis of the alignment and of the works with respect to the historical and archeological context*

Coming from the East, the alignment runs parallel with the coastline in an area that was created by the lava flow of 1669. The analysis of archeological evidence in this area didn't reveal sites or assets of any significant importance. The area is located certainly much farther East than the eastern limit of the ancient Greek and Roman urban findings. The level of archeological risk in this context is therefore deemed 'low'.

West of the port area there is the new station, "Duomo/Porto", located slightly more to the North with respect to the existing railway line.

The twin bore tunnels start from Duomo/Porto station and cross through Piazza Currò, close to the *Terme dell'Indirizzo* complex, without interfere with the archeological remains. The tunnel overburden (about 13 m) suggests that the excavation does not intercept ancient settlements levels and, as proven by the boreholes data, that it crosses through the morphological high existing already in Roman times.

Figure 6 shows the superposition of the project structure on the geological profile and of the boreholes S1, S2, S3 and S4. The 16th century town wall intercepted by the borehole S3 is highlighted and it is located above the tunnel crown.

Furthermore, the borehole S4 shows that there are no archeological elements at depths more than -6.70 m. Boreholes S1 and S2, drilled further South with respect to the *Terme dell'Indirizzo* area, confirm the change in depth of the roof of ancient alluvial deposits located above the Pleistocene clays. Most probably, the lower depths relate to the ancient levels of human activity of the port or to paleochannels relating to the ancient waterways that cross through this portion of ancient Catania – as shown in ancient maps, from North to South and into the sea.

More to the West, the area enclosed by the 16th century walls and that most of the ancient Greek, Roman and Medieval findings, becomes significant from an archeological and historical viewpoint. The analysis of these remains, in fact, shows that the tunnel bores through an area abounding with remains, with direct planimetric connections between archeological and project environments.

The bored tunnel section that is to run inside the 16th century town walls crosses through Pleistocene deposits and, therefore, poses no archeological risk.

Outside the perimeter of the walls, S-W, the strata of lava dating back to 1669 run. This condition has obviously generated an absence of archeological data which without the

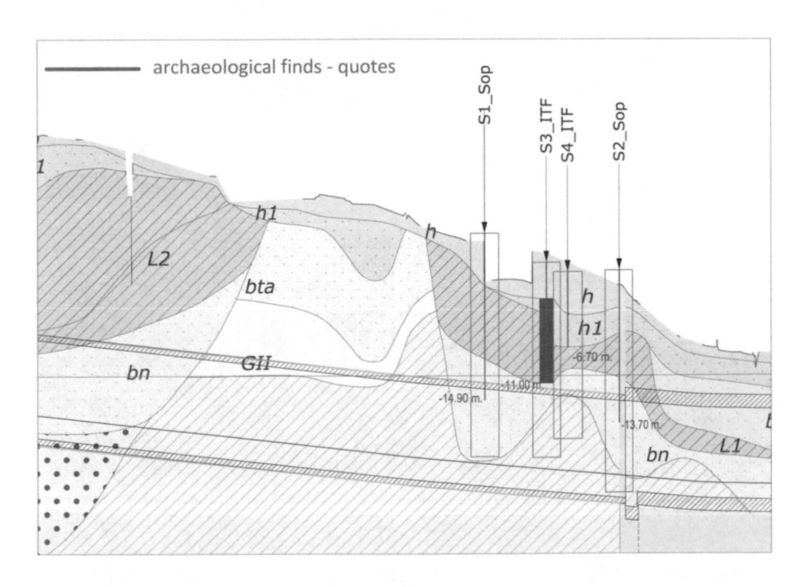

Figure 6. Superposition of the boreholes S on the geological profile.

eruption would be located under the layer of lava. However, this is an area situated next to the south boundary of the first Greek and Roman urban settlements.

The only element worth mentioning is located just west of the town walls, where the axis of the bored tunnel follows a path that intercepts the hypothetical position of the Roman circus. The earliest descriptions of this building date back to 17th century records that place it outside the *Porta di Decima* gate. This hypothesis was then mapped in 1873 by A. Holm. The existence of this building has been accepted over time a number of scholars despite the confused description of historians and the uncertainty of its architectural appearance. Considering these factors and the considerable depth of the project, a 'null' level of archeological risk has been assumed for the section of tunnel that would interfere with this alleged circus.

For all of the remaining parts of the project to the West, all the way to the Catania Fontanarossa station, the level of archeological risk is defined as 'low'.

5 CONSTRUCTION ASPECTS OF THE S. CRISTOFORO TUNNEL

The tunnels design had to combine the requirements typical of underground works with those for the preservation of the city's archeological and historical heritage.

Twin bore single-track tunnels start from new Duomo station and end to Acquicella station. The planimetric and altimetric alignment layout of tunnels has been checked and updated continuously with the geological, historical and archeological surveys above mentioned, particularly in the areas of the *Terme dell'Indirizzo* and of Castel Ursino.

In the ancient Thermae area, the alignment has been moved as far away as possible from the Roman public baths in order to eliminate any risk of settlement caused by the excavation. Figure 7 shows the Peck's subsidence curves not considering for safety reasons the favorable contribution of the layer of lava. The curves show the absence of subsidence despite the high values of volume loss (VL = 1%), for k values of 0.35 and 0.5.

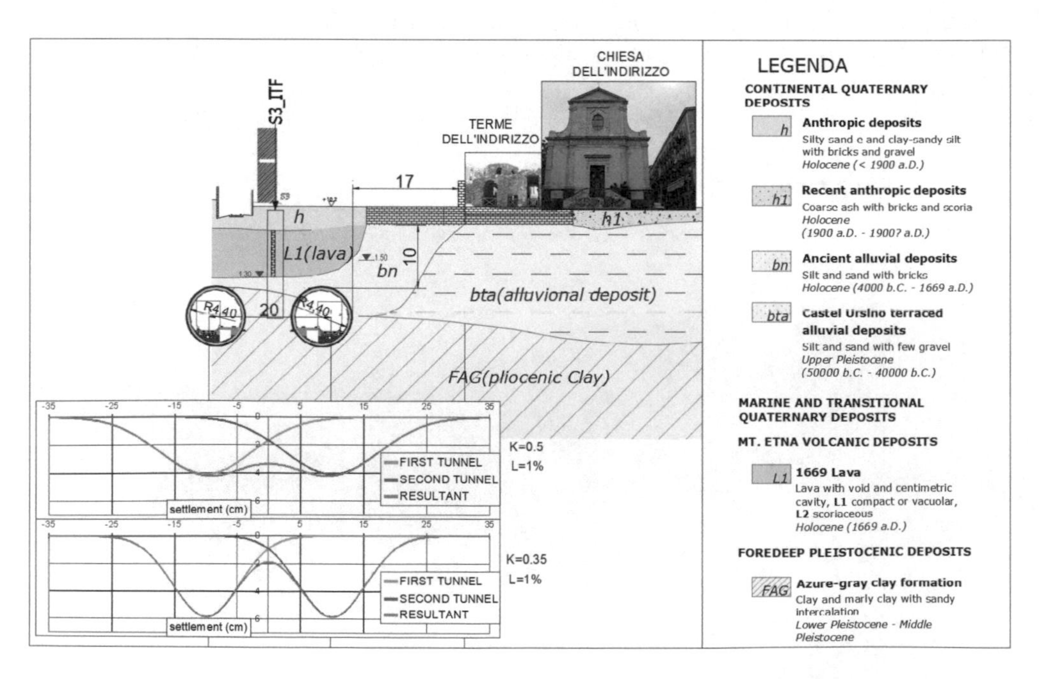

Figure 7. Geological cross-section with transvers settlement troughs in the Terme dell'Indirizzo baths structure.

The railway gradient has been designed with the maximum inclination such to ensure the greatest overburden under the buildings located between Piazza di Castel Ursino and the Duomo station.

The two tunnels shall be bored using EPB – TBM and have an internal radius of 4.40 m (Figure 8). The radius is wider than standard solution (R = 4.0 m), in order to have a 3.5 m wide platform serving the S. Cristoforo station (Figure 8).

Therefore, the choice of tunnel diameter makes tunneling phase independent from the station construction phase.

The two tunnels will be bored using only one TBM always starting from the launching shaft in Acquicella station area and to retrieval shaft in Duomo station area. After the completion of the first tunnel, the TBM will be retrieved from the 'Duomo' shaft and placed back in 'Acquicella' shaft.

On one third of the alignment buildings with low overburden and highly deformable soil below the water table are located. These are favorable conditions for the earth pressure balance technology application. However, the presence of highly resistant lava rock strata at the tunnel face makes difficult the progress of excavation, despite of the disk cutters placed on the TBM cutter head. The worst tunneling conditions occur when there are layers characterized by different strength (soil/lava) at the tunnel face, with the lava rock layer in the center of it, and become critical when there are buildings with low overburden and shallow soil of poor quality. In these conditions, the cutters positioned in the central area of the cutter head are subjected to 'skipping' along the line of stratigraphic change between soil and lava, and this can cause faster wear. The presence of lava layers also hinders a proper soil conditioning. For this reason, the twin bore tunnels have been shortened in order to avoid the tunneling in these geological critical conditions. Figures 6 and 9 show that at the retrieval shaft the lava layer remains above the tunnel section. The stratigraphic conditions in the TBM exit area vary quickly also transversally to the alignment, which means that the two tunnels are in very different geotechnical conditions even though the distance between tunnels is less than 20 m. In this area, in fact, it is located the boundary of the lava flow occurred in 1669 that, after by-passing the Castel Ursino morphological high, deviated from the sea towards the city (Figures 1 and 7).

The presence of a just few meters thick layer of lava between the crown of the tunnel and the foundation of the building is important because it reduces the entity of settlement and distortions induced by tunneling. For this reason, in the final design phase it will be necessary carrying out a 3D stratigraphy model below buildings which are underpass by tunnels with low overburden.

Another peculiar feature of the project is the interference of the tunnel with the underground river called '*Amenano*'. The hydrography of the city of Catania changed with the eruption of the 1669 that obstructed the rivers mouths in the area next to the new Duomo station. The tunnel is located just below the sea level and this should avoid any direct interference with the river. Surveys and more detailed studies are under way aimed at reconstructing the underground fluvial net.

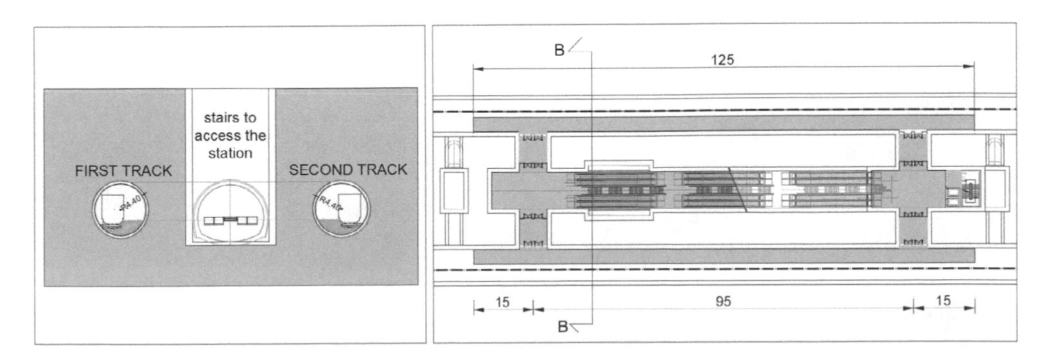

Figure 8. Cross-section and floor plan of the new S. Cristoforo station.

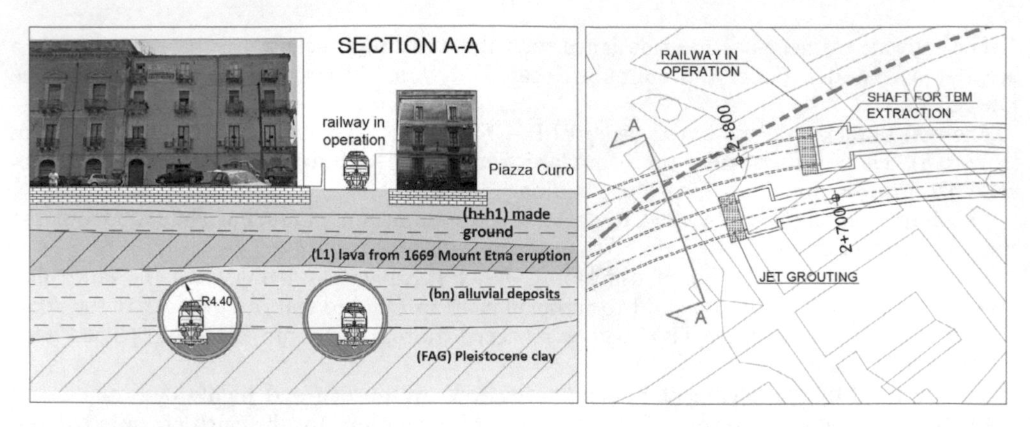

Figure 9. Cross-section closed to TBM extraction shaft.

6 CONCLUSIONS

For the new underground railway line between Catania Centrale station and Catania Acquicella station, geological and archeological studies were conducted in order to limit the risk that the works could interfere with the rich historical and archeological heritage of the city of Catania. The investigations carried out in the area and the preliminary studies excluded any significant direct interference of the twin bore tunnels with archeological heritage and allowed to obtain the approval of the public authorities (Soprintendenza). The design layouts of the tunnels alignment and of the retrieval shafts have taken into account the marked stratigraphic discontinuity between soil and lava rock, which could impair the TBM tunneling progress.

REFERENCES

Ardito F. & Scifo A. 2002. Catania archeologica. Catania

Branciforti M. G. La Rosa V. 2010, Tra lava e mare. Contributi all'archailoghia di Catania. Catania: Le Nuove Muse

Branciforti M. G. 2013. Le Terme dell'Indirizzo di Catania. Regione Siciliana. Palermo

Nicoletti F. 2015. Catania Antica. Nuove prospettive di ricerca. Regione Siciliana. Palermo

ISPRA, 2009. Carta Geologica d'Italia alla scala 1: 50000.Foglio 625, Acireale. Servizio Geologico d'Italia

Monaco C., Tortorici L., 1999. Carta Geologica dell'Area Urbana di Catania. Scala 1:10.000. S.EL.CA., Firenze.

Tunnels and Underground Cities: Engineering and Innovation meet Archaeology, Architecture and Art, Volume 1: Archeology, Architecture and Art in underground construction – Peila, Viggiani & Celestino (Eds)
© 2020 Taylor & Francis Group, London, ISBN 978-0-367-46574-2

A workflow process for tunnels maintenance. The case of the Construction Method developed for Rhaetian Railways (UNESCO World Heritage Site)

F. Modetta, A. Arigoni, S. Saviani, K. Grossauer & M. Hohermuth
Amberg Engineering AG, Regensdorf-Watt, Switzerland

ABSTRACT: All the underground infrastructures require important financial investment and if they are not adequately maintained could produce more costly and extensive repairs of tunnels. The standardisation of maintenance strategy usually foreseen for serviceability and safety reasons can reduce the life cycle maintenance cost ensuring the requested level of structural safety.

In the first part an integrated process starting from the preparation phase and using an integrated approach with traditional inspection in field and digital inspection developed to define the tunnel status assessment applying statistical analysis and numerical evaluations is described.

In the second part the selection of the most effective maintenance methods using a standardized workflow is explained to reduce the maintenance costs and to guarantee tunnel safety maximizing the quality of the tunnel repairs in presence of structural degradation. Different methods commonly applied in Switzerland are summarised and particularly the applied "Standard Construction Method" of Rhaetian Railways (UNESCO World Heritage Sites) is described.

1 INTRODUCTION AND OBJECTIVES

All the underground infrastructures require important financial investment and if they are not adequately designed and maintained follow-up costs could be higher than the construction costs (Figure 1). Early decisions have a significant impact on the costs during the operation phase, nevertheless an adequate maintenance strategy is fundamental in order to reduce life cycle costs ensuring the requested level of safety and the required level of performance.

In Switzerland in 2017 the railway network had approx. 560 single or double tracks tunnels in operation with a total length of 813 km (approx. 80% of them were built more than 50 years ago). The national roads network had approx. 360 tunnels with a total length of 420 km (approx. 20% of them were built more than 50 years ago). At this moment, other approx. 60 tunnels (around 90 km) are under construction or only planned.

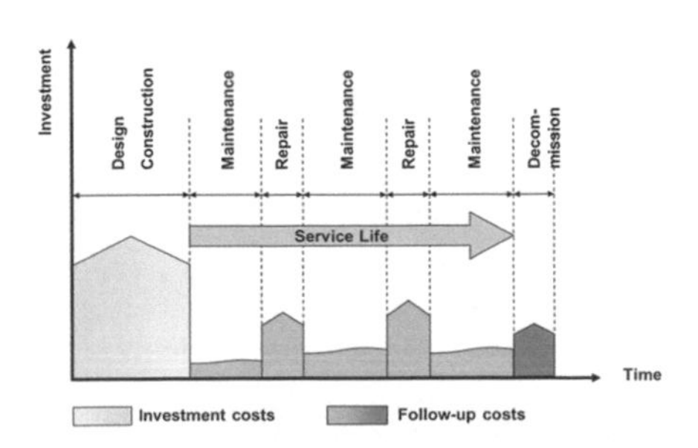

Figure 1. Overview investments scheme.

For the development, operation and maintenance of infrastructure facilities for rails and roads, several million Swiss francs are spent every year on tunnels. According to the market study InfraSuisse, the investment in 2017 was more than 1200 Mio CHF and although the investment requirement has a strongly decreasing tendency (up to 90%), it will still amount to a total of approximately 300 Mio CHF (50 Mio CHF for railway tunnels, 252 Mio CHF for roads tunnels) in 2023. Within these amounts a 85% is planned for maintenance of roads tunnels and 65% for maintenance of railway tunnels.

Within this framework, in order to use the limited financial resources efficiently, solutions that consider cost-effective construction methods with regard to operation and life-cycle costs are needed. In addition to the requirements to build both cheaper and faster, also the operating and maintenance costs are to be kept as low as possible. Furthermore, aspects such as maintenance under operation with as little impairment of use as possible must be taken into account in conservation projects.

In this context, where funding is limited and life cycle engineering is getting more and more important, an optimised and standard procedure for maintenance design becomes a fundamental aspect for underground facilities.

2 LIFE CYCLE ENGINEERING WITH ATTENTION TO FINANCIAL RESOURCES

Main goal of refurbishment projects is the restoration of the structural safety and of the serviceability for a defined time duration.

There are several reasons why structural materials of existing tunnels must be maintained and renewed. The most obvious is that the deficient condition of structural materials (e.g. severe cracking or spalling) often necessitates construction measures in order to maintain structural safety. Another common goal of renewing structural materials may be to guarantee operational safety and to reduce the required amount of maintenance. A typical construction measure for operational purposes includes the locally improving of the drainage to prevent damages to railway installations or the protrusion of ice into the tunnel. Furthermore, maintenance measures also have a preventive character against advancing damage to structural materials.

Examples of structural measures, that could be adopted locally or in the entire tunnel, are:

- Reinforcement of the lining/Replacement of construction materials in the lining,
- Repair/Renewal of joints (work joints in concrete, stone or masonry joints),
- Repair/Renewal of the tunnel drainage and/or waterproofing measures in the lining,
- Repair/Renewal of the tunnel invert. Etc.

Figure 2. Example of structural measures.

New operational requirements or the necessity to increase the level of safety could lead to the implementation of modification measures. Occasionally increased clearance demands necessitate construction measures (for more powerful and larger locomotives, panorama vehicles and wheeled-vehicle transports that require a larger clearance envelope). Besides higher speeds require a correction of the track position or an optimization of the geometrical alignment. Additional utilities may also require more space for installations, for example for installing smoke extraction systems or ventilation fans. The fires in tunnels in the last years (e.g. Mont Blanc, Tauern, Kaprun, Gotthard road tunnel) increased tunnel safety awareness and motivated operators and authorities to increase the amount of self-rescue equipment in tunnels (installations of escape paths, fire-resistant lighting and handrails require specific construction measures).

Examples of modifications of the structures are the following:

- Profile enlargement (local or extended areas for increased space demands – larger rolling stock, electrification – or capacity demands),
- Redesign and reinforcement of load-bearing elements, due for example to changes in the load conditions on the structure or new regulation requirements,
- Removing of a structure from operation by demolition or permanent filling, for new or different utilization of a structure,
- Improvement of escape and safety equipment and facilities (emergency exit, protective structures, ..). Etc.

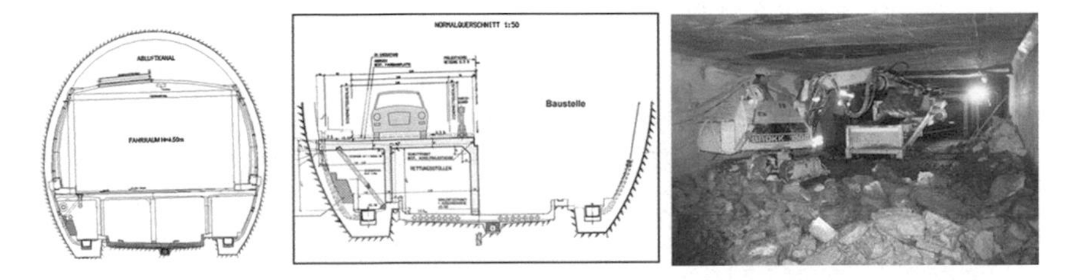

Figure 3. San Bernardino Tunnels: execution of works in the invert under traffic conditions.

The importance of systematic monitoring and periodic interventions is clearly shown in Figure 4. Periodic inspections and maintenance interventions result in lower overall costs (grey line), allowing the owner/operators to *act*. On the contrary, the lack of regular inspection and maintenance interventions lead to higher investment costs (black line). Without inspections, the owner/operators are forced to *react*.

Hence, the cornerstone of a sound maintenance management is knowing the tunnel conditions at any time: systematic condition assessment is the basis for any maintenance design and therefore for the successful implementation of construction measures.

Hereafter, the most effective maintenance methods using a standardized workflow are introduced. This process allows to reduce the maintenance costs and to guarantee tunnel safety maximizing the quality of the tunnel repairs in presence of structural degradation.

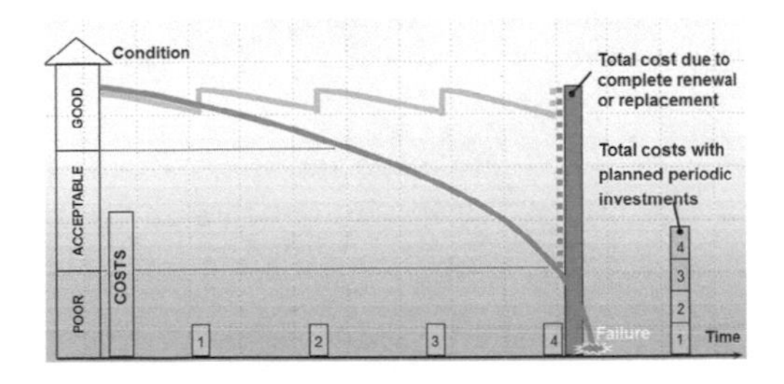

Figure 4. Systematic monitoring and interventions.

3 THE WORKFLOW PROCESS IN TUNNELS MAINTENANCE

The condition of a tunnel must be known to develop an effective maintenance concept. As the knowledge of the current condition increases in detail, the more precisely an evaluation of measures can be carried out. Only the detailed knowledge about the condition yields the basis for a focused management process of all necessary measures for structural value preservation.

Knowledge about the tunnel state is obtained in two steps. The first step consists in a comprehensive collection of technical data, which concern the properties, the characteristics and the condition of the tunnel including damages (*monitoring phase*). Then the tunnel specialists carry out the evaluation and interpretation of the data in a second step (*state assessment*).

The condition assessment can bring out the necessity of implementing different measures and activities (*interventions design*), that can be effectively designed basing on the information provided by the interpreted results of the investigation phase.

In the standards of the different countries quite different terms for structures preservation and maintenance are defined, whereby all standards express the same content: preservation is divided into survey measures and maintenance measures. In the following these aspects are detailed.

3.1 *Monitoring*

The goals of the monitoring and of the state assessment are therefore to determine the current and future condition of a structure to create a basis of information for the planning of the works.

In particular, the survey measures serve both as initial documentation of the condition and as continuous or periodic documentation of the condition development.

First of all, the following preparatory tasks are accomplished:

- Determination of monitoring goals,
- Study of existing data and documents about the tunnel (design, ..),
- Development of monitoring concept,
- Organization and coordination of the activities to be carried out.

Then the real activities can start. The characteristics and condition of the tunnel are investigated; for example:

- Construction materials and their main properties,
- Shape, internal profile and geometry of the tunnel,
- Damages and deficiencies (water infiltration, cracks, segregation of aggregates, joint damages, deformations, carbonization, corrosion, fire damage, lime deposits, ..),
- Installations for operation and clearance profile. Etc.

Additional data and indications should be collected concerning the factors which could cause and influence this condition (subsequently called the influence factors); for example:

- Construction method and tunnel age,
- Geology and geotechnics, with the related "geologically induced" forces,
- Hydrogeology, water presence and water chemistry,
- Climatic conditions (temperature, ..),
- Rolling stock and operation types. Etc.

The survey measures are subdivided into inspections, visual and measurement control, and function and material tests. Two types of inspection methodologies can be adopted:

- Traditional inspection methods based on grid cells analysis (i.e. use of manual techniques of observation and recording by checklists),
- Digital inspection technologies, such as tunnel surface mapping and profile scanning, profile and/or track measurement (Figure 5, Figure 6).

Nowadays the tunnel engineer disposes over numerous computer-aided tools for carrying out both the survey activities and then the data analysis and high-sophisticated inspection software tools support the experienced engineers in collecting the data in an optimized and economic way. A combination of the laser scanning and inspections is always necessary to integrate the results and verify any possible mistake.

In Switzerland, the minimum time interval between inspections campaigns is 5 years. In fact, the benefits of a systematic inspection process are undeniable:

Figure 5. Tunnel surface analysis and profile scanning examples.

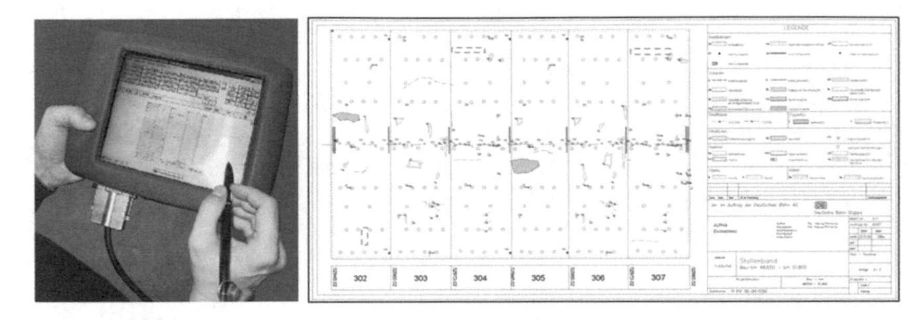

Figure 6. Tunnel Mapping.

- Minimising the interruptions in operation,
- Increasing safety and serviceability,
- Creating reliable infrastructure documentation,
- Reducing time consuming field work (minimising traffic interruption),
- Lowering the overall follow-up costs.

3.2 *State assessment*

After having collected all the necessary data, the statistical analysis of the results (quantity, distribution, comparison of damages), the damage analysis and the analysis of structural safety can be carried out in order to develop an effective intervention concept. The main tasks of the state assessment phase are indeed:

- Analysis of results obtained from observations, measurements and investigations;
- Preparation of clear documentation to show the results.

The structural elements of a tunnel as well as the installations (track, signals, etc.) are analysed but also other external factors (geology and hydrology), clearance requirements of rolling stock and compliance with current regulations shall be integrated into the state assessment.

The typical content of the output documents includes:

- Quantification of recorded phenomena and presentation of the investigations results,
- Assessment of the damages and their level,
- Rating of structural elements with analysis of structural safety,
- Recommendation of immediate measures to be taken,
- Recommendation of further procedures and measures to be implemented.

The last step now under development is the preparation of a reliable and automatic tunnel scanning and data processing tool with a complete output by a 3D BIM model. The basic workflow stays identical as mentioned above. The main goal of the ongoing studies is to make the whole process easier, more automated and therefore less time consuming and cheaper as well.

The focus is on the data processing and data analysis itself. There are many ways for collecting data efficiently in automatic or semi-automatic way whereas the inspection and damage drawing itself has always been a tedious and time consuming manual work. The workflow optimization can be done at two levels.

The first step is helping the operator doing manual drawing by highlighting the areas with high crack probability, warning him to focus more in detail in the problematic zones identified.

The second one is an automatic phenomena representation: damages are found and drawn fully automatically and the operator can then do just a visual check, saving up to ¾ of inspection data processing time.

The final goal is to reduce manual work in tunnel inspections as much as possible. Nowadays it does not seem possible to skip manual input and expert knowledge completely. Experts still need to decide about some unclear cases, but at least automatic crack record is on the horizon.

Not only the inspection itself is the issue. Another time-consuming part is the result evaluation and data transfer. The goal in this field is to be fully BIM compatible. The Amberg Tunnel Cloud already combines position information (3D), feature info (1D), time (1D) and powerful data analysis. The missing part of the chain, under implementation, is a BIM format that allows easy transfer to other programs.

3.3 *Maintenance and/or modification measures*

Whenever the assessed conditions of the facilities are such that structural and/or operational safety is endangered, changes in operational requirements are needed for serviceability reasons, safety and/or operational requirements are not fulfilled anymore, specific measures shall be implemented.

In fact, generally speaking, maintenance measures can be subdivided into preventive maintenance, corrective maintenance and renewal/modification measures.

After having clearly identified the objectives basing on the previous assessments and results, the possible technical and design solutions for the execution of the interventions are evaluated. Then, taking into account also the numerous interactions between construction, operation and environment, the best ones in terms of effectiveness, quality, time and costs are selected.

For instance, the overall costs can usually be increased due to:

- Construction phases defined based on small stages to ensure safety,
- Special (for tasks or sizes) and expensive equipment for the construction activities,
- Short production working time (with more time necessary for the construction),
- Extra pay for nights shifts,
- Huge amount of safety measures during construction. Etc..

At the same time specific choices can decrease the overall costs, by minimising the impact on the traffic, minimising the loss of income for the railway (and maybe also for roads) company and eliminating the loss of customers due to malfunctioning of the transport network.

It is clear that experience and expertise are essential in order to define the most object-specific and economic solutions. Systematic condition assessment is the basis for any maintenance design and therefore for the successful implementation of construction measures.

Besides, tunnels operators and owners are responsible for the structural and operational safety of the tunnel facilities. The implementation of a sound maintenance strategy shall be mandatory and moreover on site daily decisions must be taken in order to avoid/reduce any risks.

4 EXAMPLES OF PROJECTS ON RAILWAY TUNNELS IN OPERATION

Magnacun (1909 m) and Tasna (2351 m) tunnels, in the Rhaetian Railway in Lower Engadine, were built at the beginning of the 20th century. The first tunnel is in a formation which consists mainly of slate with gypsum layers, gneiss and amphibole. The second one lies geologically in an extremely difficult terrain of slates with gypsum and a thrust fault. The structures are built entirely of broken and cut natural stone masonry. When the maintenance was decided, the tunnel linings were deeply saturated in some sections, with the joint mortar heavily deteriorated due to frost effects, the masonry bonds at the abutments were locally loosened, with cracks and missing stones. Aggressive groundwater was the cause of local deterioration of shotcrete and corrosion of the rails. The clearance conditions were insufficient in some areas.

The concept was selected to minimize immediate costs: only specific sections of both tunnels were chosen for interventions during rail operations. Therefore work in the tunnels could only be carried out in night shifts; travel to the tunnel portals was only possible by rail such as all supply and waste disposal and all the activities could be carried out from the trains. Two different trains were used (one for sealing and shotcrete, one for demolition and support works).

Since the works were carried out during the night, a clear division between construction and railway operations existed. It must be taken into consideration that the works were carried out in max. nine hours at night (including travels to the site and clearing of the site).

Due to the chosen installations concept, it was possible to carry out the works without disturbing the normal rail service and within the budget (ca. 1850 Euro/m).

Emmersberg is a 761m long tunnel built in 1895 on the railway line Etzwilen-Schaffhausen (Swiss Federal Railways). It is in a solid moraine and in partially saturated water-bearing sand and it has an elliptical profile lined with natural stone masonry made of cut limestone. Exploratory drillings and maintenance works in the invert showed that no substructure had been built and the ballast was separated from the rock surface only by an approx. 5 cm thick sand layer.

The 20–25 cm thick ballast layer was generally too shallow, this resulted in a poor stability of the track system, which suffered local settling of the rails. Besides groundwater could not be effectively drained as a result of sand deposits in the drainage system: water backed up in the tunnel invert which caused a saturation of the underground, resulting in mud deposits and ultimately the instability of the track system. Points of water infiltration were observed in the side walls and in the crown, reducing the durability (corrosive effects of track supports and the rails).

The clearance as well as the track system and substructure problematics were solved through renewal of the tunnel invert including a lowering of the track level. The works were carried out in phases. Firstly preparations and works in the vault were carried out during night in six hours shifts for nine weeks. Then an eighteen weeks interruption in rail operations was planned for works in the tunnel invert. After putting the tunnel back into service, the rest of the works in the tunnel vault was planned to be completed during night shifts for seven weeks. In a specific stretch to avoid buildings settlements, the invert excavation length and the concreting stages were reduced and to provide extra support steel braces were applied in each excavation stage.

The total construction costs in the most critical area were approx. 7500 Euro/m.

5 CASE STUDY – THE RHAETIAN RAILWAYS

A particular example is the Rhaetian Railways (included in UNESCO World Heritage Sites). The Rhaetian Railways AG (RhB) operates an approx. 384 km long rail network, predominantly in the Canton of Graubünden (CH). With the rail network situated in Alpine area, there is an extensive use of tunnels (115 tunnels, total length of 58.7 km, constructed between 1901–1914).

Due to the tunnels age, more than half of them require renovation (approx. 75 tunnels, for approx. 26 km) within the next 50 years. A new rehabilitation strategy was necessary to ensure that the existing rail tunnels satisfy current tunnel design standards with the primary goal to ensure that renovation costs are lower compared to previous renovations by following a standardized process during the tunnels conditions assessments and also during construction.

Based on the workflow above described, the primary challenges associated with the renovations concern the determination of the extent of repairs as well as creating a feasible concept, which allows a new requirement (tunnel maintenance during operations).

The tunnels are single-track rail mostly lined with masonry, some of them were also put into operation without linings. The tunnels show considerable damages to the load-bearing structures, mainly due to wetting and frost. The water behind the lining and the associated formation of ice lead to damage to the masonry joints and to high lateral pressures with high deformations particularly in the side walls where, together with the structurally unfavourable profile (straight sides), they lead to systematic weak points in the tunnels. In addition to the weakening of the load-bearing structure, sudden breaking out of material near the tracks is an increased operational risk, which could lead to train derailments.

5.1 *Repair Concept of 2010*

The first concept for tunnel repair was developed in 2010–2013 (Argenteri, Charnadüra and Klosters Tunnels). It is based on the idea of holding the penetrating groundwater with a 5 cm thick shotcrete layer and draining it with local drainages. In order to ease the water behind the masonry, weep holes are drilled at regular spacings; the existing deformation is only excavated locally and the tunnel lining is rebuilt. To maintain the required clearance profile with special values, the tunnel invert is lowered and rebuilt with a concrete invert and sometimes the widening of the tunnel sides is carried out .

Permanent waterproofing against wetting and damage to the masonry cannot be guaranteed with a 5 cm shotcrete layer. The analysis showed that these repair works (localised joint repairs, application of shotcrete sealing, without tunnel invert) have a cost of about 10,000 CHF/m but the lifetime can reach maximum 25 years; with some other activities, using the same 2010 concept the costs arrive to about 22,500–30,000 CHF/m still assuring only a maximum time horizon of 25 to 50 years. In comparison to this the costs for a new tunnel are about 55,000–65,000 CHF/m with a lifetime of 100 years. Hence, a new concept with a time horizon at least >50 years is evident.

5.2 *Standard Tunnel Construction Method, Study of 2012 & development*

The aim of the method developed in the study 2012 was to solve the most frequent causes of damages in almost the tunnels, lowering the investment and achieving a repair interval of 70–100 years in addition to the following requirements:

- The tunnels should be repaired during operations,
- The weak point of tunnel sections (straight side walls) should be remedied,
- The installation of an invert should prevent heave and softening,
- An improvement of tunnel safety should be achieved,
- The tunnel profile should correspond to the specification of the clearance profile.

In the first concept developed by the RhB with Amberg Engineering and F. Preisig AG, the existing masonry side walls were modified on both sides (along about 20 m to 30 m each side was enlarged, temporarily supported and replaced by precast concrete elements). The use of precast elements also in the tunnel vault was regarded too critical since the loose rock and the risk of collapse of the unknown backfilling behind the masonry. For this reason, the solution proposed was to apply waterproofing to the masonry, held in place by a 15 cm thick shotcrete layer supported on the prefabricated side elements. The tunnel invert was built with cast in place concrete invert to ensure the ring closure, increasing the resistance also to horizontal forces compared to the original concept, and to drain the invert (Figure 7). The final adaptation of the standard clearance profile gauge resulted in track lowering by 70 cm.

To evaluate the feasibility, a test tunnel about 40 m long (Figure 8) was excavated at the test tunnel in Hagerbach, with the same profile of the RhB tunnels, to verify the handling of 1.5m long precast elements and to analyse the time required to set the segments. The test with the setting of 12 precast elements in 6 hours was done by three contractors with different methodologies: the time for the trial assembly was reduced by all considerably (to 120–150 min).

The following critical issues were highlighted after the 1:1 trial:

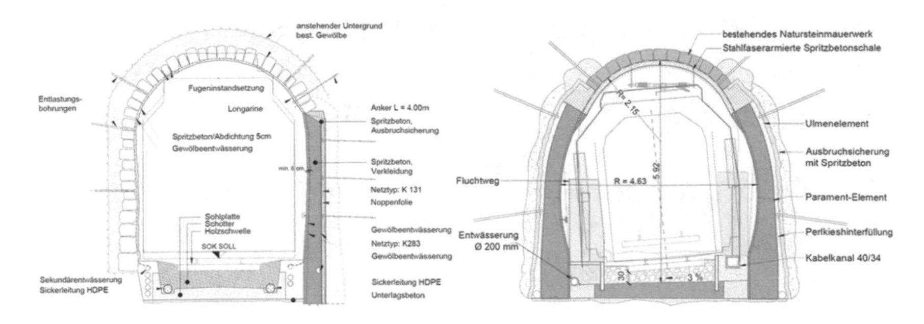

Figure 7. Left: Repair concept of 2010, tunnel profile with replacement of the local side wall and right: Repair concept as per the study of 2012 (Design by Amberg, Grüner, Basler&Hoffman).

- The transition from the precast elements to shotcrete was a structural special solution,
- Due to the material technology, there was a discontinuity at the transition, the long-term effects of which cannot be quantified,
- On various projects such as those with adjacent bridges, the necessary track lowering before and after the tunnel was impossible due to constraints.
- The temporary support and underpinning of the existing masonry vault along a length of 20 to 30 m was regarded as an obstruction to the construction.

After the successful test assembly and for the above listed reasons, the final design was adapted:

- Replacement of the guiding rod with a plinth to fix and adjust the side element,
- Use of a precast crown element instead of leaving the masonry crown.
- Gaining space "upward" reprofiling the crown instead of lowering the alignment,
- With the crown element, the geometry of the side elements was rethought. From two original elements planned 19 standard elements were developed (Figure 8).

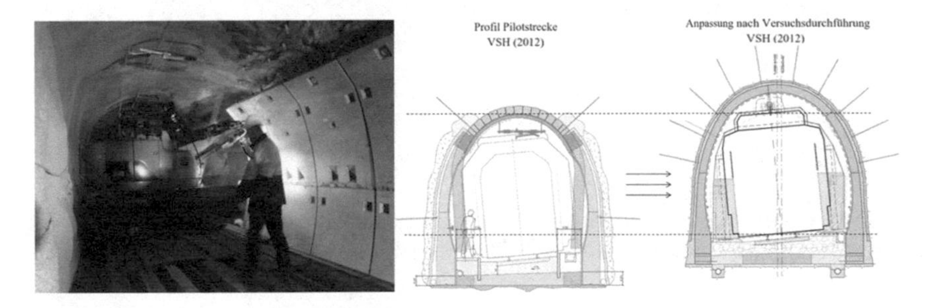

Figure 8. Left: The test gallery after the placement of the precast side wall elements. Right: Initial profile used for testing in VSH & final profile for the Glatscherastunnel (Design as in Figure 7).

A concept for working during operations near the overhead had also to be developed. The solution was a mobile protective lining, which supports the masonry with hydraulic cylinders and protects the critical zone behind the face from falling masonry. A telescoping overhead rail enables the excavation of the crown without overhead. Figure 9 shows the position under rail operation and the position during the excavation of the tunnel crown.

5.3 *Standard Tunnel Construction Method Tunnel, Implementation at the Glatscherastunnel*

This method was applied in the approx. 334 m long Glatscheras Tunnel. The standard profile is a closed ring of precast concrete elements (length 1.5 m, thickness 30 cm). Since the crown was enlarged, vertical track level adaptation could be omitted and the new invert was changed in 1.2 m long poured concrete slab. After the setting of the invert, the trapezoidal lower and

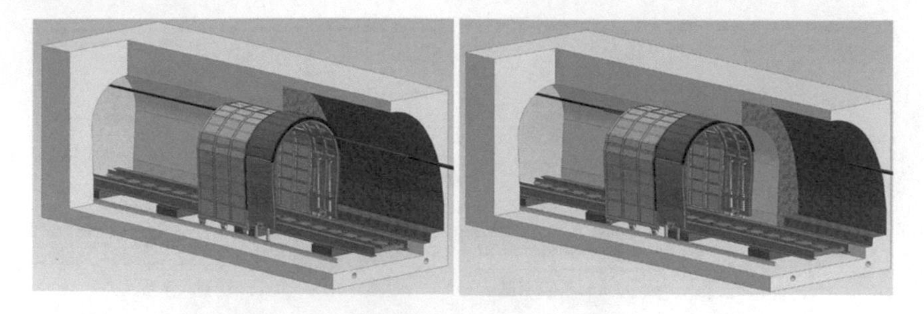

Figure 9. Position of the protective lining during operation (left) and construction (right) (Herrenknecht).

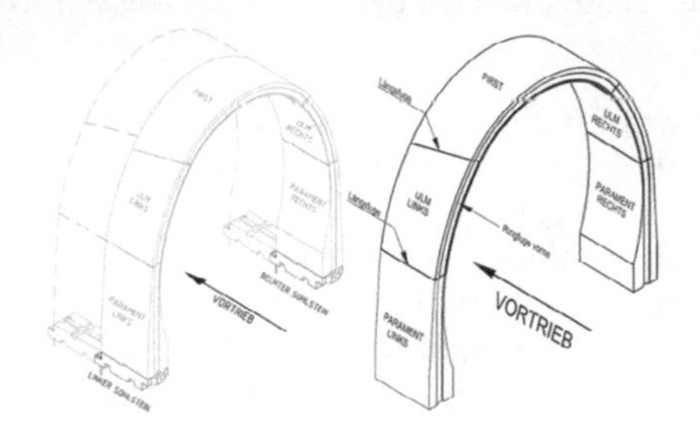

Figure 10. The proposed placements of the precast concrete elements (Design as in Figure 7).

the upper sides elements were set; finally a precast crown element was provided. The longitudinal joints have flat contact surfaces sealed with rubber gaskets and similarly the ring joints are sealed with a special profile; bolted mechanical connections were used. The surface of the invert elements was formed horizontally in the longitudinal direction with a guiding rod. The ring joints were formed with a special wide guide groove and equipped with a joint profile on the rock side.

This construction increases the lifetime up to 70–100 years and will permit the RhB to repair one or two tunnels per year lowering the cost to 42,500 CHF/m for future repairs.

6 CONCLUSIONS

The workflow usually applied in the tunnels assessment and in the maintenance works design demonstrated to provide always more detailed and reliable studies and solutions. Besides an organised approach based on repetitive processes and on past similar experiences, together with a detailed tunnel assessment before any measure definition, have two major direct results: engineering time and costs savings and at the same time construction costs savings in terms of direct cost and of reduction of claims during construction (due to sound, optimised and feasible solutions). About the "Standard tunnel construction method, study 2012" concept, the standardisation of the construction process for maintenance allows contractors to adapt equipment and machinery to a method that will provide further time and cost savings for the clients.

REFERENCES

Amberg Engineering AG 2012. *Dokumentation Durchführung* VersuchsStollen Hagerbach AG.

Baumann, K. 2014. *Normalbauweise Tunnels Konzeptbeschrieb*, Rhätische Bahn Infrastruktur.

Design documents by Amberg Engineering AG, Gruner AG and Basler & Hoffman AG.

Grossauer, K. Modetta, F. & Tanner, U. 2017. *The "Standard Tunnel Construction Method" of Rhaetian Railways*. In OGG Geomechanics and tunnelling, vol. 10. Berlin: Ernst&Sohn.

Herrenknecht 2013. *Machbarkeitsstudie Schutzkonstruktion*.

InfraSuisse 2017.

Tunnels and Underground Cities: Engineering and Innovation meet Archaeology, Architecture and Art, Volume 1: Archeology, Architecture and Art in underground construction – Peila, Viggiani & Celestino (Eds)
© 2020 Taylor & Francis Group, London, ISBN 978-0-367-46574-2

The First World War military tunnels of the Italian-Austrian front

S. Pedemonte
Military History Enthusiast

E.M. Pizzarotti
Pro Iter, Milan, Italy

ABSTRACT: During World War I, particularly in the years 1915–1917, all along the Austrian-Italian Front spreading from eastern Lombardy to the Gulf of Trieste, many intense tunnel works were on going for diverse military purposes. Some of these underground activities, many of which are nowadays well preserved and can be visited thanks to the constant effort in conservation and restoration by the local Authorities, are impressive if one thinks of the difficult environmental conditions in which they were built and the technical challenges of the operations. A remarkable literature exists on this subject from the historical and biographical point of view. This paper, instead, focuses on technical and technological aspects, on material resources, manpower, design and construction means and methods used to build tunnels that today seem extraordinary in relation to the period in which they were completed and to the difficulties encountered during the execution.

1 INTRODUCTION

It cannot be denied that he inspiration to deal with this topic stems from having read two interesting articles appeared on Tunnels&Tunnelling (O'Reilly 2016 & Backhouse 2017), on the tunnels excavated during WWI on the Franco-German front of the Marne. The tunnels of the Austrian-Italian front are no less significant and deserve to be brought to the attention of the technical audience. Unlike those of the Marne front, these tunnels were built in a mountainous territory, often at altitudes above 2000 m a.s.l. and therefore in logistical and environmental conditions decidedly more difficult (think about the extremely snowy winters back then, with very rigid temperatures). Furthermore, the characteristics of the materials involved in the excavations were completely different, given the generally competent rock masses, requiring the use of explosives (so the advancement could be easily heard by the enemy – Pasquale in press). Their construction covers a period that extends from a few years before the declaration of war (24 May 1915), for the preemptive preparation of defensive fortifications, until the end of war operations (4 November 1918) and their geographical location involves the whole front from eastern Lombardy to the Gulf of Trieste (Fig. 1) and extends further west to include the Ticinese region. This paper focuses on the underground works of most relevant interest completed in the period 1915–1917 on the eastern-central portion of the front itself, bounded on the west by the Adige valley and on the east by the Tagliamento river (Dolomites region). In this sector, with few exceptions, the opposing armies have mostly exploited the underground not only for purposes of logistics (deposits, walkways, shelters, etc.), but also as a real offensive tactical tool. The most articulate and demanding works are represented in fact by the Mine tunnels, used to reach a point below enemy defensive positions and create a breach by blasting huge quantities of explosives. Finally, this essay will be limited only to the works carried out by the Italian Army, even if the Austrian ones are certainly not less important.

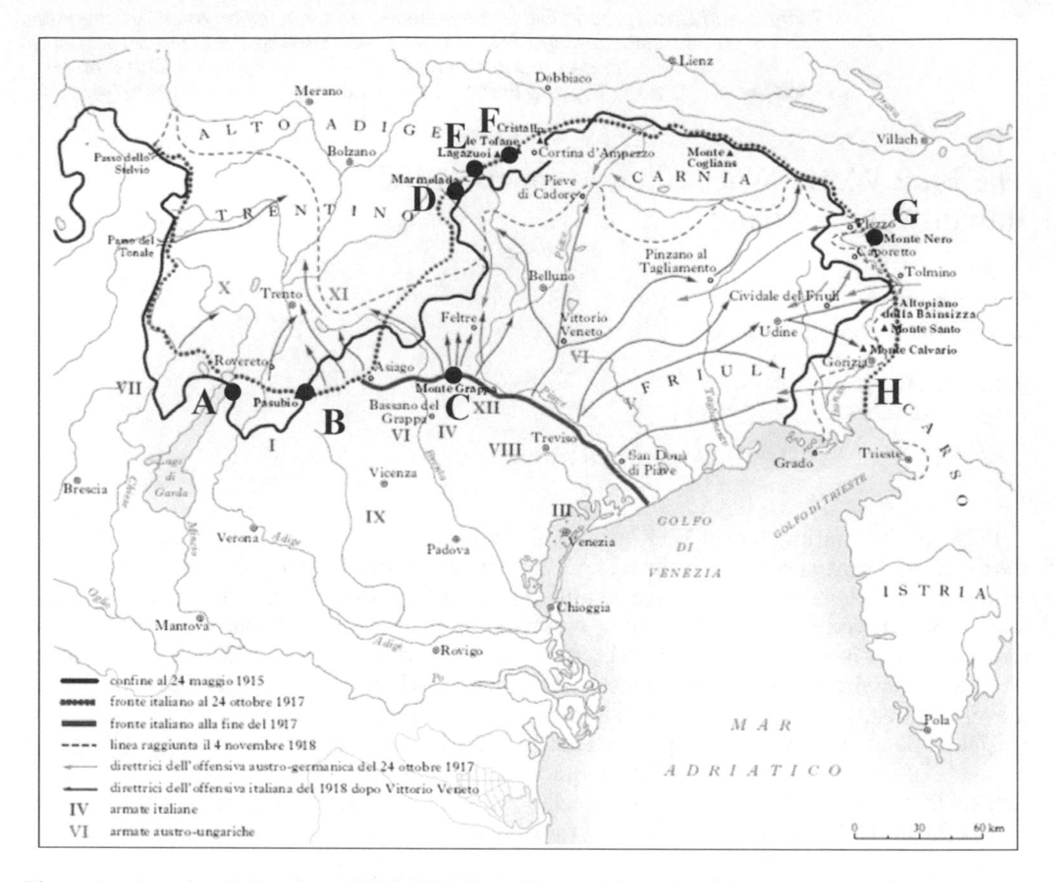

Figure 1. Austrian-Italian front 1915–1918 (from Treccani Encyclopaedia). A: Monte Zugna. B: Pasubio. C: Monte Grappa. D: Marmolada. E: Cima Col di Lana. F: Lagazuoi and Castelletto di Rozes. G: Catena Monte Rosso - Monte Nero. H: Carso.

2 GEOGRAPHIC AND GEOLOGICAL FRAMEWORK

Figure 1 shows that at the opening of warfare the active front covered over 800 km. The underground fortifications extended further to the west in the Ticinese and in the upper Adda valley since an invasion by the Austro-German armies was feared across Switzerland. Excluding this western sector, Italy began to fight against the Austro-Hungarian Empire on a front of about 500 kilometers (Touring Club Italiano 1928) running from the border with Switzerland to the Gulf of Trieste and mostly mountainous with alpine stretches in some cases reaching 3200 m a.s.l. (Adamello massif).

As shown in Figure 2, the area is entirely included in the Southern Alps south of the Periadriatic Fault. It is a tectonic unit with characteristics clearly different from those of the other Alpine structural units. The schistose-crystalline rocks, mostly phyllite, are common only in the basement, while the powerful sedimentary caps, mostly from the Mesozoic age, are dominant. The covering layers are small, whereas in the western part décollements and slidings are frequent. The structural movement is given by a succession of folds, fold-nappes, and tilted scales, mostly verging to the south. There are impressive magmatic masses both intrusive, like the Adamello tonalitic mass, and extrusive like the powerful castings of Permian Quartz-porphyry of the Porphyric Atesina Platform (Desio 1973). Information about the rocks crossed by the main underground operations carried out during war are provided below, some of which will be briefly outlined thereinafter, proceeding from west to east (Fig 1).

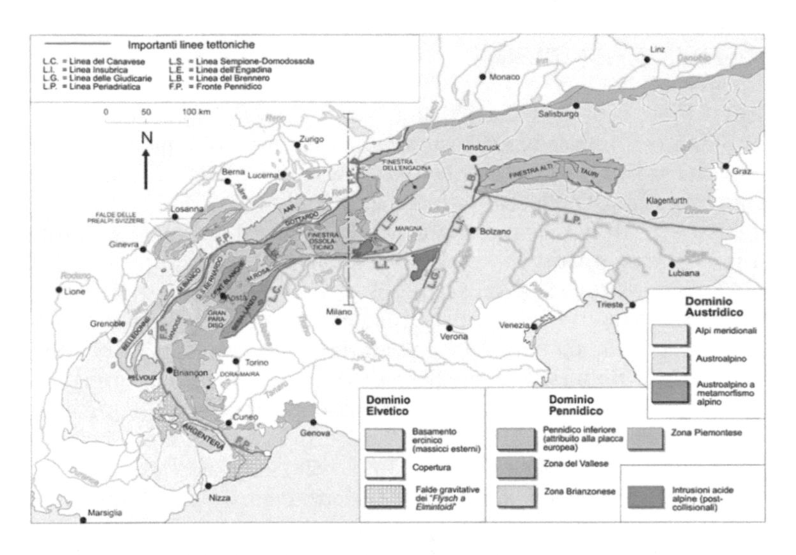

Figure 2. Tectonic Scheme of the Alps (by Daniele Fornasero).

A. Monte Zugna - The underground works cross metric layers having a structure composed of dolomite and limestone rocks referable to the Dolomia Principale and the Grey Limestone Unit, which are often subject to important karst phenomena.
B. Pasubio - The Pasubio is a carbonate, dolomitic and, in the upper parts, also partly calcareous massif: the tunnels therefore cross both limestone and dolomites. The formation characterizing the area is the Dolomia Principale: dolomite sediments of carbonate platform, reaching over eight hundred meters in thickness, crystal-dolomite rocks in large banks, alternated with micritic dolomites (fine-grained) and stromatolite dolomites.
C. Monte Grappa - Sedimentary organogenic carbonate rocks, formed by accumulation of plants and shallow sea organisms (limestone): they are compact and fine-grained. The units crossed are those of the Gray Limestones. Some outcrops confirm a non-massive but largely grainy texture, with a rock that is not very compact and tends to flake apart.
D. Marmolada - Here we have massive gray limestones, debris and stromatolites. It is the Marmolada Limestone Formation, resulting from coral cliffs, at whose bed appears at times the Livinallongo Formation composed of nodular and tuffite limestones, also found in the area of Col di Lana, described below.
E. Col di Lana mountain - In the area affected by the tunnels, there is the Wengen Formation and, more occasionally, the Volcanoclastite Submarine Landslide Accumulations Unit. The Wengen Formation includes mixed volcanic carbonate (tuffite) deposits with massive texture. Immediately in the bed of these two formations there is the Livinallongo Formation often heterotopical to the Sciliar Massif.
F. Lagazuoi and Castelletto di Rozes - The tunnels are located in the well-layered light gray Dolomia Principale, having at its base silts, shales, and multicolored marlstones of the Raibl Formation and further underneath the crystal and massive Cassiano Dolomites.
G. Monte Rosso - Monte Nero Chain (Krn – currently in Slovenian territory) - Massive limestones, with regular and massive layers, sometimes in banks. Among the crossed formations it is worth mentioning the Scaglia Rossa (sedimentary fossiliferous marine rock, siliceous-calcareous lithology) and the Dachstein Limestone, the latter outcropping on Monte Krn (Monte Nero) and almost similar in its composition to the Grey Limestones of Monte Grappa.
H. Carso - Here we have white bioclastic limestones, massive or rubbles with dolomite clasts as well as whitish stratified limestones with residual clays. The main outcrops are the Liburnian Formation, the Aurisina Limestones, the Monrupino Formation, and the Monte Coste Limestones. There is a widespread and imposing speleogenesis.

3 THE FORCES IN THE FIELD

The peculiarity of WWI was the presence of a huge number of men and vehicles close to the front, barely moving and exposed to the shooting of cannons and machine guns as never seen before. The defense works carried out before 1915 were punctual and spread over a few stretches, ending up for not being useful to the economy of the conflict: many were located at the border of Switzerland and France anticipating different strategic purposes. The Italian Army faced the first war operations mobilizing 31,000 officers, 1,058,000 troopers, 11,000 civilians and 216,000 quadrupeds (Ministry of War 1927). At the end of the war, these numbers reached three times higher values. In particular, the Military Engineers rose from about 12,000 units in 1915 to 110,000 (equal to one ninth of the Infantry) in 1917, reaching at the end of the war 170,000 mobilized (equivalent to one fifth of the Infantry). No other corps underwent a similar increase, a clear sign of the importance of engineering and, above all, of underground works. For example, before war the 5th Specialist Miners Regiment had 13 units and in 1918 ended with 53 (Touring Club Italiano 1928), so it is reasonable to assume about 11,000 men, plus the *centuriae* (civilians directly dependent on the Army) and private companies still working behind the lines. An approximate reckoning of workers, involved by the Military Enigneers but not part of its troops, was of 110,000 men in 1917 for the *centuriae* and 650,000 civilian workers throughout the war. During operations a sub-specialty "Motorists" of the 5th Military Enigneers Regiment was created along with a test center in Milan in contact with the national supplier of drilling groups. It was possible to produce less unwieldy, lighter, and safer equipment compared to those received from abroad or previously produced in Italy. A type of motor-compressor gave excellent results and was also adopted by the French and English Armies. Each Italian Army set up a repair workshop with the task of training personnel (Touring Club Italiano 1928). These data give the idea of the organizational effort then carried out by the Italian Army: from isolated and local episodes at the beginning of the war, the tactics of underground warfare became general and organized at the end of the war, with part of the Italian industry that specifically produced new explosives, drillers, electricity generator, pumps, fuses, and anything else that was required. The experience was gradually spread through memos, manuals and books, involving also the Italian mining schools.

4 THE WORKS

It is almost impossible, at present, to quantify analytically the action of the Italian Army in the field of underground works. However, we will try to give, through the most striking examples, a picture of the effort made. It is on the Carsot, sadly known for the continuous frontal attacks, that the Italian Army faced the new type of war in July 1915 and considered the opportunity to resort to underground Mine works (Di Martino 2017). But everything remained hypothetical because of the lack of skilled labor and technical means until January 1916, when the High Command gave broad and prevalent development to Mine works for offensive purposes on the heights of Santa Lucia di Tolmino at the base of Mount Krn, of Podgora in front of Gorizia and of San Michele del Carso. It should be noted that the High Command ordered the requisition of materials suitable for that purpose from firms operating in the mining sector only then and gave the Army Technical Office a mandate to make a census of the existing equipment at national level. A first project concerned Cima 4 of San Michele del Carso (Fig. 3a) where two 80 m tunnels had chambers loaded each with 125 kg of explosives. Many other proposals were made, but the works proceeded slowly even for the lack of specialized men: consider that the 1st Unit of the 5th Specialist Miners Regiment did not have the necessary fleet and only 10 soldiers were miners by profession (Di Martino 2017).

On the Monte Zugna (up to 2000 m a.s.l.), the Mines and Counter-mines warfare continued until September 1918 through wells and tunnels described "in fissured rocks". The following data on the excavations as of December 31, 1917 are given as an example: a steep slope tunnel of 20 m for hearing out the Austrian works; a shaft of eight meters from which a 14 m tunnel and another 24 m tunnel with a 25% gradient start; a great number of natural cavities came upon during the works (Bertè 2013).

On the Pasubio (2239 m a.s.l.) (Figs 3b, 4a), around 10 km of tunnels and 500 km of roads were built, 1500 explosive charges were burned every day, 100 km of compressed air pipes and 60 km of water pipes were laid (Làstrico 1940). Among others, the works included also a 110 m long helical tunnel with a 2.2 × 2.5 m section, a 190 m long tunnel with a 2.0 × 3.0 m section and another one of 140 m with a 2.5 × 2.5 m section. Excavations were carried out with an average production of 6 m^3 per day. 5 hammer drills were used, while for electricity there was a Ballot generator together with a portable generator for ventilation (Di Martino 2017). But this massif is famous in particular for the "Road of 52 tunnels" which starts at 1216 m a.s.l. and ends at 1928 m a.s.l. with a total of 2335 m of tunnels along a 6300 m long path (Gattera 1995).

On the Monte Grappa (1775 m a.s.l.) there is the "Vittorio Emanuele III" tunnel system (Fig. b), built in just 10 months, which has a main branch of 1400 m and a total development of

(a) (b)

Figure 3. (a) The tunnels of San Michele del Carso. (b) Tunnel "Generale Papa" at the Pasubio. (authors' photos).

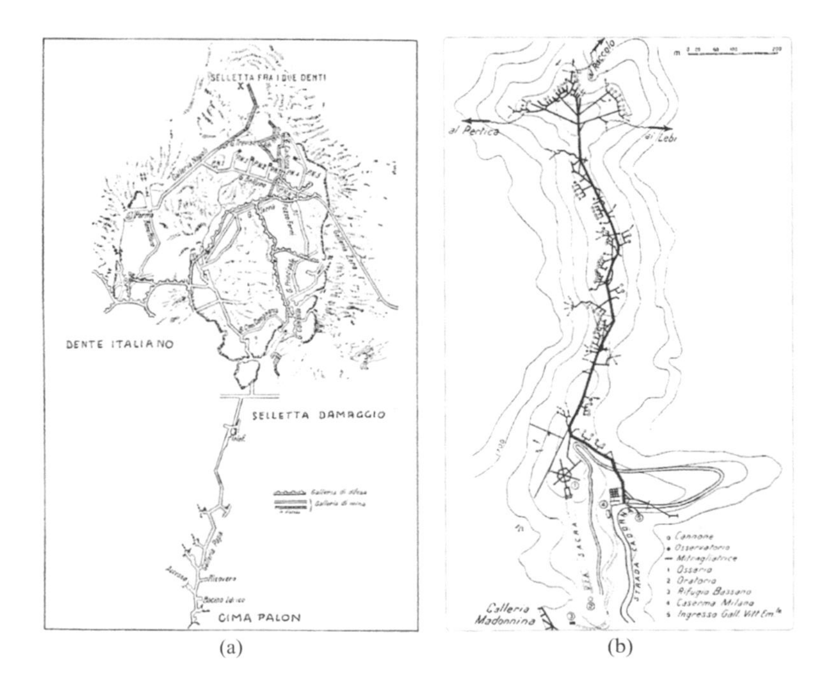

(a) (b)

Figure 4. (a) Network of Italian tunnels on the Pasubio (Pieropan 1990). (b) The complex of tunnels of Monte Grappa (Touring Club Italiano 1928).

5152 m: it housed 23 artilleries of which six 105 mm cannons, plus machine guns, food storages, water and ammunitions (Touring Club Italiano 1928). It could house 1,500 soldiers and the deposits with 50 to 200 m^3 water tanks guaranteed 15 days of endurance (Massignani & Bellò 2001). It is worth noting (Fig. 1) that Monte Grappa is much further south of the front line at the beginning of the war. The defensive system was built well before the advance of the Austrian Army in October 1917 and, together with the Piave River, it represented a strategic bastion that proved to be insurmountable allowing to stop the said advance.

The Marmolada is famous for the tunnels excavated within the glacier, although even here, at 3065 m asl, the military engineers of the two armed forces had to excavate hundreds of meters of rock tunnels: the "Rosso" Tunnel can still be visited today. (Fig. 5a).

The Piccolo Lagazuoi (Fig. 5b, 6a) was not only the theater of Alpine warfare, but also of daring engineering achievements. Consider that to load the Mine with 33 t of explosives, our soldiers dug a helical tunnel of about 1100 m with a difference in height of 250 m and a gradient up to 60% that required the use of steps (reaching 2660 m a.s.l.). The tunnel was 1.90 m high and equally large and the excavations advanced at a rate of 5.5 m/day (Di Martino 2017). Here the engineers did not use dynamite for charging the Mine, but gelatin and Echo.

At Castelletto (2657 m a.s.l.) of the Tofana di Rozes (Figs 6b, 7a) a 500 meters tunnel was excavated in a very steep slant for 2200 m^3 of total rock, with a large Mine chamber in which 35 t of explosive were blasted on 11 July 1916 (Pieri 1968). With regard to the shooting of enemy lines on the top of Castelletto, the tunnel adit (Figs 6c, 7b) was placed in a sheltered position within a natural ravine and required the overcoming of significant logistical difficulties.

It is still possible to admire the complex of Italian and Austrian tunnels in the Julian Alps starting from Pontebba, Val Dogna, Rombòn, Monte Canin up to Monte Rosso, near Monte

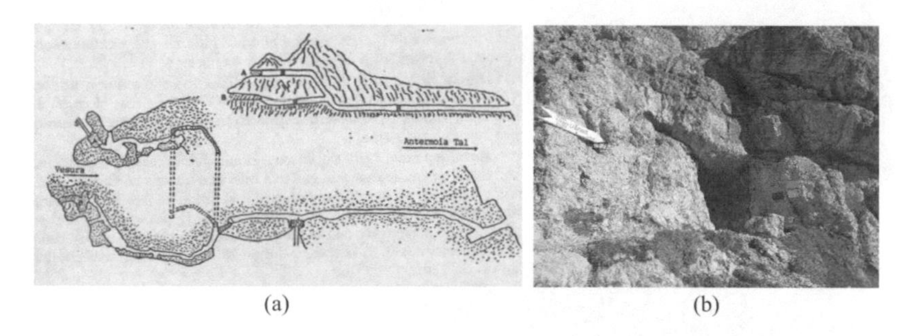

(a) (b)

Figure 5. (a) A sketch of the "Rosso" Tunnel, whose entrance was freed from the ice a few years ago (Di Martino 2017). (b) Entrance of Lagazuoi tunnel (authors' photo).

(a) (b) (c)

Figure 6. (a) Lagazuoi tunnel. (b) Castelletto tunnel. (c) Access to Castelletto tunnel. (authors' photos).

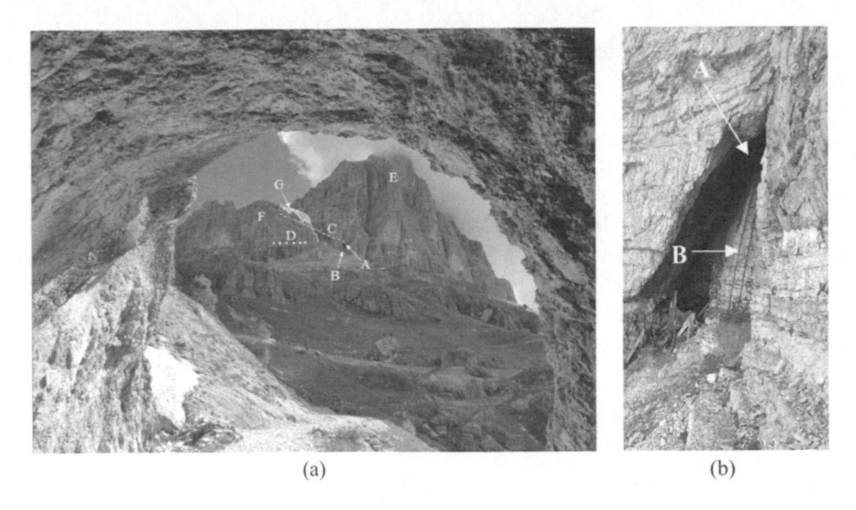

Figure 7. Castelletto Mine tunnel, Tofana di Rozes: (a) Overview from a tunnel placed below. (b) Adit. (authors' photos). A = Tunnel adit (Fig. 6c), B = Ladder to tunnel adit, C = Castelletto Tunnel (Fig. 6b), D = Austrian Embrasures, E = Tofana di Rozes, F = Castelletto, G = Ridge destroyed by Italian Mine.

Nero (Krn), of which memorial articles tells us of incoherent and very fissured rock that required a great job of reinforcement; not far away, the excavation on Mount Mrzli also encountered a crumbling mass.

As previously mentioned, the Mine tunnels, are among the most complex and demanding works carried out. Table 1 summarizes the main data of the most important Italian Mines and Counter-mines. We must not forget that the opponent was equally busy digging tunnels and caverns: to mention are the Mines at Monte Cimone (14 t), the one at the Lagazuoi (24 t) and that at Monte Sief with 45 t of explosives exceeded only by the record 50 t of the Pasubio Mine.

As we have seen, the rock masses involving underground works built by the Italian Army are of discrete, and often good, geomechanical properties. The reports regarding the construction of these tunnels, either for shelters, warehouses, or Mine are limited to the threats caused by the enemy with the Counter-mine tunnels. The rock characteristics or the need for stabilization operations are almost never mentioned: an obstacle was rather the mechanical resistance to the drilling slowing down works.

Table 1. Some of the Italian Mines and Counter-mines.

Date	Type	Location	Amount of Explosive [kg]
17/4/1916	Mine	Col di Lana	5000
12 in 1916	Counter-mine	Carso	
11/7/1916	Mine	Castelletto	35,000
31/7/1916	Counter-mine	M. Mrzli	
1917	2 Counter-mines	M. Sief	3600 and 5000
1917	3 Mines	Colbricon	
8/6/1917	Mine	M. Zebio	1000
10/6/1917	Mine	M. Rotondo	
23/6/1917	Mine	Lagazuoi	33,000
1917	2 Mines	M. Mrzli	
16/8/1917	2 Mines	M. Rosso	
1917	2 Mines	Pasubio	13,000 and 1000
1917	2 Mines	Marmolada	500 and 1000
1918	3 Counter-mines	Pasubio	1200 in one
1918	2 Counter-mines	M. Zugna	825 and 2100

A particular case is the tunnel dug in the bed of the Popena torrent, in the Cristallo group: in 1916 units of the 54th infantry regiment dug a tunnel of about 350 m that was partially under the bed of the river (Di Martino 2017) with a 10 meters cover: the speed of advancement, despite the dripping and the conditions of inconsistency of the front, was of two meters in twenty-four hours. The coating consisted of 6 cm thick planks and the excavation section reached 2.2 × 2.0 m. But on the 19th of June the entrance caved-in blocking the tunnel, while on the 9th of July great parts of the tunnel were destroyed by the opposing artillery causing it to be abandoned.

The underground environments visited by the authors (tunnels, culverts, shelters, road tunnels, caves for shelters or artillery) at the Pasubio, Lagazuoi, Castelletto, Monte Mrzli (close to Monte Nero), Monte San Michele and Carso, are not lined and still almost undamaged after 100 years. Only small sections had minor falls in rare, particularly fractured areas, mostly at adits.

5 TECHNOLOGICAL ASPECTS

5.1 *Explosive*

Many types of explosives were used during warfare depending on the need:

- Gunpowder (75% potassium nitrate, 13.50% charcoal and approx 11.50% sulphur);
- Dynamite (nitroglycerine);
- Explosive gelatin (92% nitroglycerine and 8% colloidon);
- Echo (60% ammonium nitrate, 25% aluminium powder, with percentages of nitrocellulose and coal);
- Cheddite (a chlorates-based mixture);
- Sabulite, nitranite or siperite (mixtures of over 70% ammonium nitrate and gunpowder).

The trigger could be made of either 1.2 gram mercury fulminate caps or pyroxylin (nitrocellulose with a high nitrogen content). For example, 1 m^3 of granite required 1 kg of gunpowder or 250 g of explosive gelatin, 1 m^3 of hard limestone 600 g of gunpowder or 180 g of explosive gelatin. The Italian monthly production of gelatin was 80 t per month in 1916 (Di Martino 2017): from Table 1 it is possible to comprehend the organizational effort made for providing the Mine of Castelletto with 35 tons of explosives.

With explicit regard to Mines, the guidelines of the 1915 Military Engineer Officer Field Manual, outlined the effect of dynamite in underground excavations, with the formula:

$$C = a \cdot m \cdot h^3 \quad (1)$$

where C = charge necessary to create a crater on the surface; a = type of explosive; m = parameter depending on lithology; h = depth of the Mine chamber. Some of the m-values are shown in Table 2. A number of other tables showed the data required for excavating tunnels.

Table 2. Some m-values taken from the *Military Engineer Officer Field Manual* 1915.

TAB. XI. — *Valori di m.*

Terra leggera	$m. =$	1,20	Murat. antica	$m =$	3,63
Sabbia forte	»	1,75	» romana	»	4,24
Terra con pietre	»	2,00	Buon calcest. di cem.	»	4,24
Argilla con tufo	»	2,25	Rocce terrose	»	2,50
Murat. cattiva	»	1,88	» compatte	»	3,27
» mediocre	»	2,42	» dure	»	4,24
» eccellente	»	3,27	» fessurate	»	5 a 6

5.2 *Drillers*

Before drillers came out, two workers were needed to open a mine hole: one holding a long steel rod and propping its end shaped for cutting, and the other hitting its head with a bat. The results were short and low charged holes. So, the Italian Army began a phase of research: a 3 cm steel pipe equipped with a diamond-drill pushed and rotated against the rock was tested; the resulting debris were removed with a water jet pumped through the core of the pipe which also served for cooling down the device. It was even possible to reach 50 m making a small Mine chamber by exploding gelatin cartridges. It was even placed a 50 kg charge at a distance of 30 m. However, this practice remained at test level (Di Martino 2017).

As for the mechanical drillers, as already mentioned, war financed the research for new technologies and here we provide a summary of some types of machines that were used:

- 45 Hp Aquila-Sullivan used at Castelletto and then at Lagazuoi;
- 30–40 Hp Sullivan used at Castelletto;
- 75 Hp Sullivan used at Lagazuoi;
- 15 Hp Alfa-Ingersoll for additional works at Lagazuoi;
- 15 Hp Diatto used at Lagazuoi;
- Romeo, Sullivan and Consolidated Pneumatics Tools, on pack animals transportable groups Marelli-Brusa.

5.3 *Advancement in tunnel*

For rock tunnels the *Military Engineer Officer Field Manual* - 1915 suggested a width of 1 meter and a height between 1.5 and 2 meters for two miners working on a 1-hour shift, or width and height between 1.8 and 2 meters for two pairs of workers. Those who visit the existing tunnels of this size may easily understand the effort that the men employed had to face because of the uncomfortable position in which they had to operate, the dripping, the dust and the bad temperatures. Some examples of daily progress are shown in Table 3.

Table 3. Advancements and tunnel's sections.

Tunnel	Advancement [m/day]	Section [m]	Notes
Monte Cimone	3.2	1.1 × 0.8	With driller
Col di Lana	1.2		By hand
Monte Sief	4.0		48 men, 4 drillers, 4 shifts
Lagazuoi	5.0 – 6.0	1.9 × 1.9	With driller.
Castelletto	5.0 – 6.0	2.0 × 1.8	120 men – 4 shifts per day
Monte Piana		2.0 × 1.8	Listening tunnel.
Monte Rosso	1.2 – 1.5		With driller
Carso	0.37 – 1.5	1.5 × 1.0 and 1.0 × 0.8	By hand.

Figure 8. The excellent characteristics of the rock mass in the access trail to Monte Cengio (Cortelletti & Acerbi 1997).

6 CONCLUSIONS

Difficult underground works were obviously carried out also in other sectors of the front (Monte Cimone, Monte Cengio, Monte Sief, Buse dell'Oro, Colbricon, etc.). With this essay we just wanted to highlight how, from the beginning until the end of war, the exploitation of underground military potentials went from limited and punctual cases to the realization of impressive and technically challenging works that required the parallel development of a real and dedicated industrial sector. The underground excavations carried out by the Italian Army during three years of war, are still largely visible and may be sightseen, so as to become a heritage not only historical but also as regards the landscape, appreciated by tourism more and more attentive and attracted by the possibility to directly verify the difficulties involved in their construction. In our opinion, such a heritage deserves greater analytical scientific commitment in terms of listing and classification.

However, we believe that there is one more reason to analyze these works from a geomechanical point of view. For example, Figure 8 is indicative of an undoubtedly excellent rock mass that in these days no engineer would ever try in a similar way: wouldn't it be useful to investigate its state of effort by means of in situ tests and to provide a back-analysis? Hundreds of kilometers of unlined tunnels are available for studying geomechanical issues. Why not transform some of these works into real geomechanical laboratories on the type of the one existing in Switzerland: The Hagerbach Test Gallery?

REFERENCES

1946. Carta Geologica d'Italia, Foglio n. 37: Bassano del Grappa. Scala 1:100,000. ISPRA.

1968. Carta Geologica d'Italia, Foglio 36: Schio. Scala 1:100,000. ISPRA.

1977. Carta Geologica d'Italia, Foglio n. 028: La Marmolada. Scl. 1:50,000. IGM. Servizio Geol. d'Italia.

2005. Carta Geologica d'Italia, Foglio n. 029: Cortina d'Ampezzo. Scala 1:50,000. APAT.

2013. Carta Geologica del Carso Classico a scala 1:10,000 – Progetto GEO-CGT. Regione Autonoma Friuli Venezia Giulia.

Backhouse A. 2017. The War Underground. *Tunnels&Tunnelling* (July).

Bertè T. 2013. *Guerra di mine sul Monte Zugna. "Trincerone" (1915–1918).* Rovereto: Museo Storico della Guerra.

Cortelletti L. & Acerbi E. 1997. *Da Cesuna al Monte Cengio.* Valdagno (VI): Gino Rossato Editore.

Desio A. (ed.) 1973. *Geologia d'Italia.* Torino: UTET.

Di Martino, B. 2017 (6th edition). *La guerra di mine sui fronti della Grande Guerra.* Valdagno (VI): Gino Rossato Editore.

Gattera C. 1995. *Il Pasubio e la strada delle 52 gallerie.* Valdagno (VI): Gino Rossato Editore.

Ispettorato Generale del Genio 1915. *Manuale per l'Ufficiale del Genio in guerra, fascicolo 1.* Torino: Tipografia Rattero.

Làstrico L. 1940. *L'Arma del Genio nella Grande Guerra 1915–1918.* Roma: Ministero della Guerra.

Massignani A. & Bellò G. 2001, *Guida al Monte Grappa.* Valdagno (VI): Gino Rossato Editore.

Ministero della Guerra 1881. *Istruzione sull'impiego della dinamite nelle mine militari.* Roma: Voghera Carlo Tipografo di S.M.

Ministero della Guerra 1881. *Istruzioni pratiche del Genio. Istruzione sui lavori da mina* Titolo II. Roma.

Ministero della Guerra 1927. *L'Esercito Italiano nella Grande Guerra, volume I (Narrazione).* Roma.

O'Reilly M. 2016. Tunnel Warfare. *Tunnels&Tunnelling* (July & August).

Pasquale E. in press. *La geologia militare e la Grande Guerra.*

Pieri P. 1968. *La nostra guerra tra le Tofane.* Vicenza: Neri Pozza.

Pieropan G. 1990. *Monte Pasubio.* Valdagno (VI): Gino Rossato Editore.

Touring Club Italiano 1928. *Sui campi di battaglia – Il Monte Grappa.*

Tunnels and Underground Cities: Engineering and Innovation meet Archaeology, Architecture and Art, Volume 1: Archeology, Architecture and Art in underground construction – Peila, Viggiani & Celestino (Eds)
 ISBN 978-0-367-46574-2

Metro Thessaloniki – intersecting microtunnels to support archeological findings at Sintrivani station

D. Rizos
OMETE S.A., Athens, Greece (currently: Orascom Constructions)

G. Vassilakopoulou & P. Foufas
OMETE S.A., Athens, Greece

G. Anagnostou
ETH, Zurich, Switzerland

ABSTRACT: This paper presents the concept, the design challenges and the risk mitigation measures of underground construction of a metro station directly underneath unexpectedly encountered archaeological findings. The construction of the station under consideration was planned to be executed with the so-called top to down construction method (cover and cut) with the use of permanent diaphragm walls and horizontal slabs supported directly by the walls. During the excavation after construction of the diaphragm walls, the lower part of a Byzantine basilica Church with an inlay mosaic floor was discovered, underneath the top slab, 5 m below the ground level. The Contractor was instructed to change the design and the construction plan in order to secure and support the significant archaeological findings. An array of six intersected tunnels of diameter about 2.5 m and length of 21.75 m was proposed to underpin the findings. The sequence included three primary and three secondary tunnels, excavated conventionally, lined with shotcrete and filled with reinforced concrete to form a permanent horizontal slab about 1.8 m below the foundation level of the findings. Afterwards the excavation continued unhindered underneath the slab up to the final level and the original construction concept applied.

1 INTRODUCTION

1.1 *General*

The case history of the present paper belongs to the first metro line of Thessaloniki, the second most populated city in Greece. The Thessaloniki metro is estimated to serve 310′000 passengers daily. The design and build contract was awarded to the JV Aegek – Impregilo – Ansaldo T.S.F – Seli – AnsaldoBreda. OMETE S.A. was the design engineer for the geological studies and the geotechnical and structural design of 13 stations, several crossovers and shafts, the 50′000 sqm large depot as well a number of conventionally constructed or cut-and-cover tunnels.

Thessaloniki Metro is a challenging project as it is being stretched in a densely urban environment with high and old buildings, including several ancient monuments that must remain undisturbed and intact from the underground construction. The very small contractually specified limits to surface settlements along with the poor geotechnical subsoil conditions and the complex hydrogeological setup in the project area comprised difficulties, which had to be carefully coped with by design and construction.

The top-down construction method with diaphragm walls was adopted for the stations, the crossings and shafts. Due to the space limitations, no inner lining could be accommodated within the stations and shafts, which means that the reinforced concrete diaphragm walls – initially acting as excavation support – represent also the only permanent vertical structure.

The high and partially confined artesian groundwater level as well as the great depth of several stations (up to 35 m below ground level) along with the low weight of the underground structures made it necessary to install a permanent micro-pile system in order to mitigate the risk of uplift. The presence of pressurized aquifer at some stations necessitated a pumping network to release the overpressure and ensure excavation stability. For one station with considerable in situ hydraulic head gradient and seepage flow downhill, the risks due to the rise of the ground water table uphill of the station were studied and mitigation measures to halt on going settlements at the surrounding buildings and monuments due to poor soil foundation conditions were taken timely (Vassilakopoulou et al. 2010).

At some stations, the insufficient passive earth resistance required deep grouting of the subsoil in order to form a bracing element by improved ground, which would be able to resist the high lateral earth and water pressure during the excavation and to reduce surface settlements to an acceptable level (Vrettos et al. 2013).

1.2 *Sintrivani station*

The station is rectangular in the plan view, 76.60 m long and 19.75 m wide. It comprises 4 underground floors, thus reaching at a depth of 27.20 m below the soil surface. Figure 1 shows a typical cross section.

The construction of the station was carried out with the top to down method, using the cover and cut construction sequence. Specifically, the latter starts with the construction of the perimetric diaphragm walls and continues with the excavation and survey of the upper archaeological soil layer. The bottom boundary of archaeological layer was estimated at level +9.94 m, which is 1.65 m below the first underground slab (slab -1). For this excavation the vertical walls were supported laterally by one row of steel struts. Then the casting of the slab -1 was executed followed by the casting of the roof slab and the backfilling above it to restore road traffic. Thereinafter, excavation took place sequentially up to the level of each intermediate slab followed by the concrete pouring of the respective slab. This sequence was followed until the raft slab. Apart from the roof slab, which rests on the walls, all the intermediate slabs and the raft slab are pinned directly to the diaphragm walls using couplers. The slab connection to the walls was structurally assessed as hinged and fully fixed connection. With this construction sequence

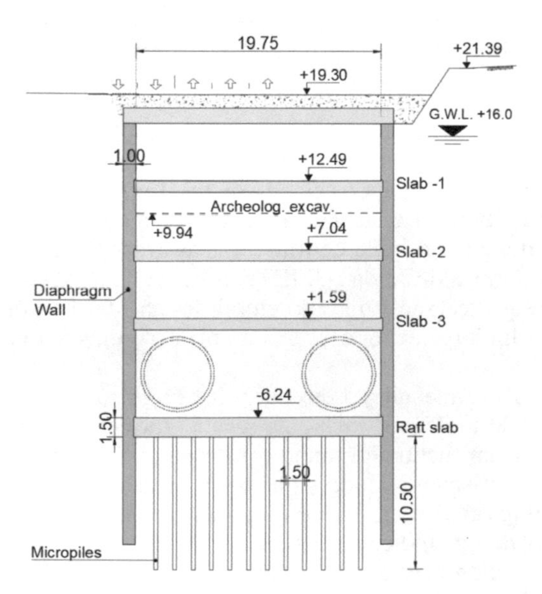

Figure 1. Typical cross section of the Sintrivani station.

the station is laterally supported by the diaphragm wall and the horizontal slabs. This so-called multi-bracing system results in the smallest possible surface disturbance, reduces settlements and environmental impact (preserving the water table).

The raft slab was anchored with corrosion-protected micro-piles, which consisted of one single rod element, had an anchorage depth of 10.5 m and were arranged in a grid of 1.50 m x 1.60 m. The safety against uplift and the design of the micro-piles were verified according to EC7 and DIN 1054. The analysis of the micro-piles was carried out considering the singe element, the group and the system structure-micro-piles.

The station box itself does not include an inner lining and hence the diaphragm walls have been designed to undertake the full hydrostatic pressure and the long-term earth pressure. The development of a high bending moment and shear force at the connection with the raft slab is avoided by the micro-piling anchoring system.

For all the accesses at the vicinity of buildings, secant pipes with anchors or struts were used to provide lateral support during the excavations of the open pits.

The archaeological survey to level +9.94 m was contractually mandatory to be carried out by the Contractor under the tight supervision and direction the 9th Ephorate of Byzantine Antiquities.

2 GEOTECHNICAL CONDITIONS

An extensive ground investigation program was executed along the route, consisting of more than 200 boreholes, geophysical surveys and pressuremeter tests. The latter served to estimate the Young's modulus of the ground and the earth pressure coefficient at rest. The performed laboratory tests included classification tests as well as direct shear, triaxial and uniaxial compression tests for the assessment of the mechanical parameters of the ground.

The survey showed that about 2/3 of the route runs through medium to stiff cohesive soils and in some areas weak rocks, while the rest consists of soft clays and silts, loose sands and gravels. At the greater area of Sintrivani station the soil profile comprises the following layers:

- Man-made deposits with variable thickness, up to a depth of 11 m (so-called archaeological layer);
- Stiff to very stiff sandy clay (CL-CI), locally with gravels of low to intermediate plasticity, with sporadic, discontinuous and thin intercalations of dense clayey sand (SC) with gravels and of clayey gravels (Layer A2b);
- Hard and locally very stiff sandy clay (CL-CI) to clay with sand of low to intermediate plasticity, to very weak claystone/siltstone, with local and discontinuous intercalations of dense to very dense clayey sand (SC) and of clayey gravels (GC) to locally of well cemented anglomerates (Layer A2c).

The ground water table was met at a depth of 2 to 6 m below the surface, while in some places artesian groundwater was found at depths greater than 25 m. Significant downhill seepage flow (associated with a hydraulic head gradient of about 1%) was detected at one station, which means that the construction of the diaphragm wall could raise the water table uphill of the station pit. Few meters away from the station an artesian aquifer was detected at depth 30 m below the surface with a piezometric head about 1 m above the natural ground water level.

3 PROJECT REQUIREMENTS

The metro line is located in a densely populated urban environment, which comprises sensitive and important buildings and monuments adjacent to the structures and above the tunnels. At the first part of the route, the buildings are founded on recent manmade deposits, including the archaeological strata. The thickness of the latter varies between 8 m and 11 m.

These conditions made the Owner to put very restrictive rules for the design and construction of the tunnels and stations. The protection of the water table during the excavation and the limitation of the settlements below 24 mm were contractual requirements to be strictly fulfilled. The excavation inside the archaeological strata and the survey were rigid conditions, which governed the technical solutions and in some cases had an unpredicted impact on the time schedule of the works.

4 ARCHEOLOGICAL FINDINGS AT SITRIVANI STATION

After completing the construction of the diaphragm walls and of the piling walls (for the excavation of the station and of the accesses, respectively), the Contractor was obliged to execute the archaeological survey under the supervision of the archaeological authorities. According to the initial estimations, the archaeological layer should be 6 m thick, but first findings during construction triggered a more extended archaeological survey, up to a depth of 11 m below the ground surface. Specifically, part of the ancient cemetery of the city was found and at the south-western part of the station important archaeological findings were revealed, about 5 m below the initial ground surface, below the planned roof slab of the station. The findings included an inlay mosaic floor (Figure 2) of an early Christian basilica dated from the end of the fourth century AD. Figure 3 shows the station layout and an aerial photograph with the position of the Byzantine basilica relatively to the station.

On account of the great importance of the mosaic floor for the history of the city, the archaeological authorities and the Ministry of Culture decided to relocate and expose it to a Museum and at the same time to cover and protect the floor of the basilica. To this end the project Owner instructed the Contractor to carry out studies for the construction of the station at the existing location without touching the foundation of the monument. The studies should foresee the excavation of part of the station under the basilica footprint. This meant that a new construction method should be developed, which takes account of the already constructed diaphragm walls, underpins the monument and makes use of the part of the station outside and below the footprint of the church.

5 UNRERPINNING

5.1 *The concept*

In fact, at the time the basilica was discovered (partially inside the station area; Figure 3), the perimetric frame of the station was already formed by the diaphragm walls. Meanwhile, the two

Figure 2. Inlay mosaic floor (courtesy of Makropoulou, 2014).

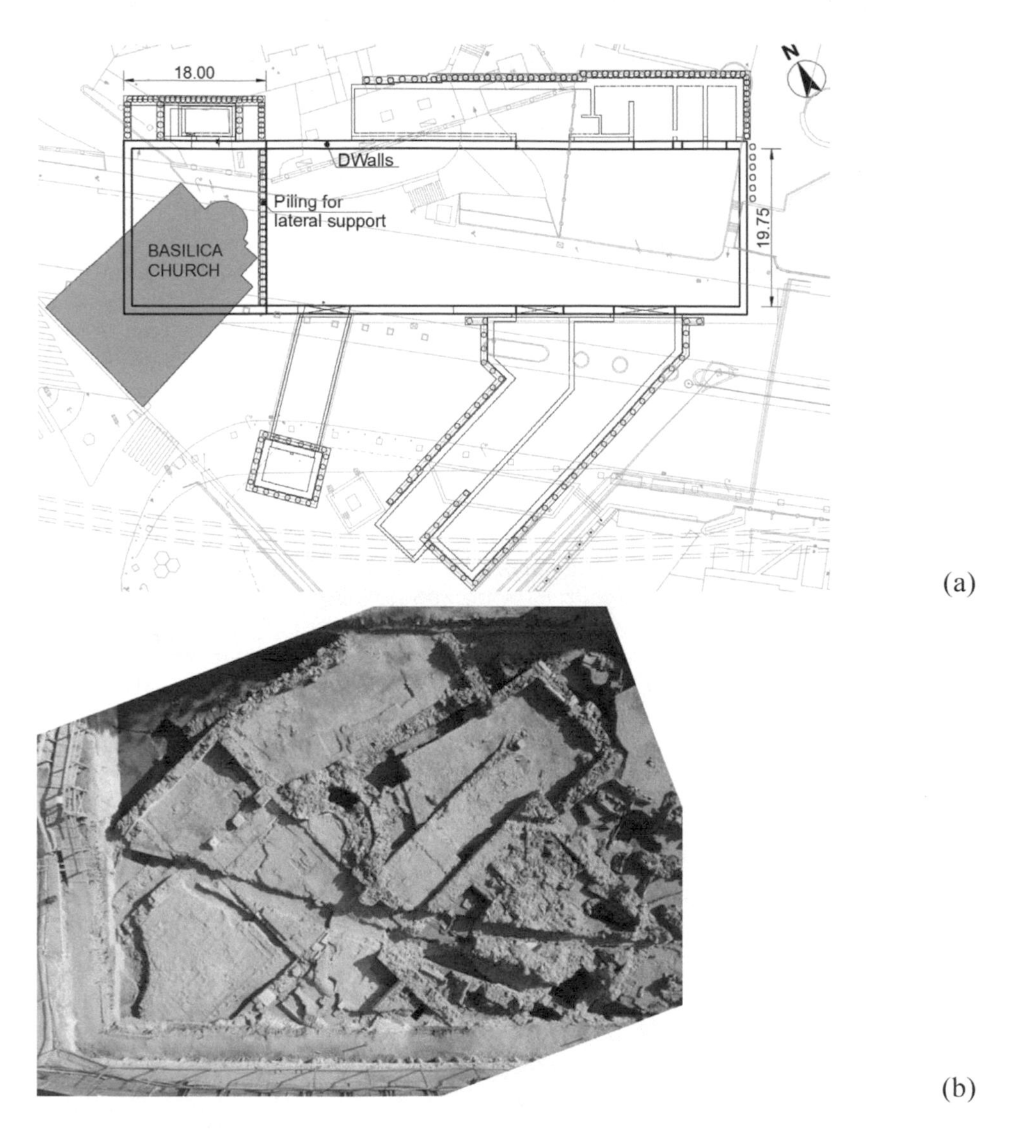

Figure 3. (a) Station layout with the Basilica footprint; (b) aerial view of the exposed Basilica inside the station (courtesy of Makropoulou, 2014).

TBMs were launched and there was no time to relocate or redesign the station from the scratch. The designer was called to optimize the design in order to support and preserve the monument, maintain the size of the station (width and length of the platforms) inside the initially foreseen boundaries, to ensure all station functionalities and the same capacity, while about 30% of the upper part of the station was not anymore available. With the new arrangement the station was divided in two parts: the part under the basilica and the part outside. Several options were analysed for the support of the basilica, considering the constraints imposed by the already constructed diaphragm walls. Actually, the latter proved to be advantageous for the support of the monument, also because the foundation of the basilica was not that much deep. The optimum solution was the underpinning of the monument by a horizontal slab, which would be supported directly on the existing diaphragm walls. This implied a horizontal underground excavation and concrete casting underneath the monument. To this end, an array of six intersecting microtunnels was chosen as the most feasible, technically acceptable and less risky solution to form the supporting slab. The six microtunnels were filled one by one with reinforced concrete to form the slab. The chosen solution was inspired by previous, successfully executed building under-crossings in Germany (Brandl, 1989) and Switzerland (Hassler, 2012).

The mean diameter of the microtunnels and their axis-to-axis distance are equal to about 2.5 m. The microtunnels were excavated all the way to the opposite diaphragm wall and part of this diaphragm wall was demolished. The one-span simply supported slab is being supported directly by the diaphragm walls. The bottom level of the microtunnels was chosen to coincide with the level of the slab level -1. This solution is structurally advantageous because the deep excavation results in similar bending moments in the diaphragm walls as initially planned and, consequently, the reinforcement of the already constructed diaphragm walls was sufficient. In addition, the elevation of the microtunnels provided a sufficient cover to ensure the safe excavation underneath the footings of the church. With this layout, only one level of the station had to be abandoned (at its western part), while the level -2 underneath the microtunnels slab was integrated in the station.

To ensure that the loads will be distributed among the individual concrete beams, which formed the slab, shear corbels were arranged at their interfaces. In addition shear dowels and reinforcement was placed to limit joint openings. Sealing hydrophilic strips were placed all the way along the contact surfaces to guarantee waterproofing at the construction joints. Injections with resin materials were conducted at the end of the slab, in the contact with the main station box. Furthermore, to ensure waterproofing under the microtunnels, a false ceiling was provided with a drainage system, which would collect any unexpectedly infiltrating water.

The six microtunnels were excavated from the northwest ventilation shaft attached to the station box. The shaft was deepen by one level and enlarged to the necessary width in order to provide the required space for the excavation of the six micro tunnels. The adopted solution and the layout of the tunnels are shown in Figure 4 (longitudinal section of the station) and Figure 5 (cross section).

A transverse secant piles wall was constructed to separate and protect the floor of the church from the excavation works at the remainder part of the station outside the footprint of the basilica. The pile wall was supported laterally by a steel bracing system, which ensured the structural integrity of the piles and of the perimetric diaphragm walls and limited the deformations during the archaeological excavation until slab -2 (elevation +7.04 m). The toe of the piles was founded sufficiently deeper than the bottom level of the slab -2 in order to provide the appropriate support for the safe excavation of the station until the slab -2.

5.2 *Construction phases*

First, the transverse secant piles beside the basilica were constructed in order to provide the necessary support for the unhindered continuation of the archaeological excavation in the open pit up to the initially foreseen level +9.94.

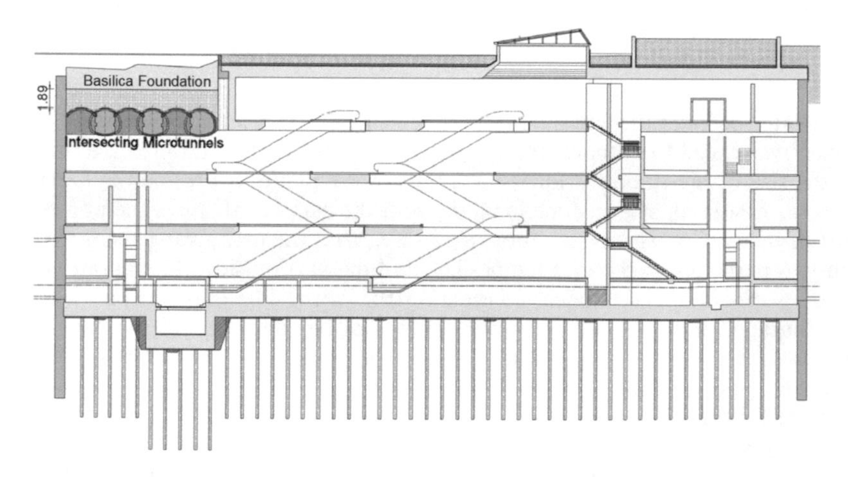

Figure 4. Longitudinal section and microtunnels layout.

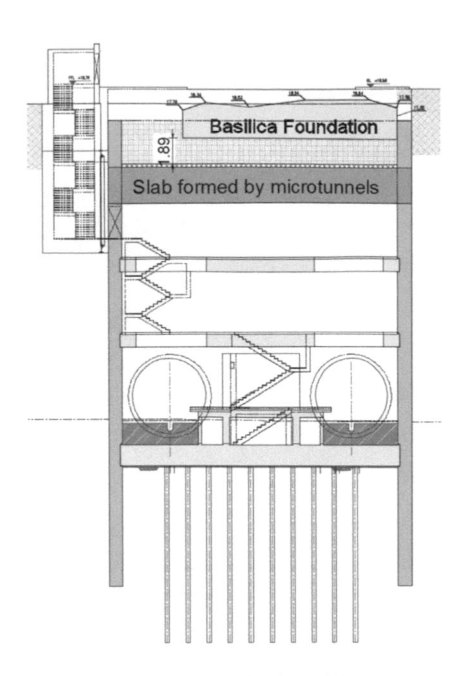

Figure 5. Cross section at the position of the underpinning slab.

Then the piles at the northwest shaft attached to the station were constructed followed by the excavation of the shaft to the bottom level of the microtunnels. The shaft was excavated later (after the microtunnels completion) to the depth of level -2 and connected with the station to provide required emergency exits.

First, the microtunnels 1, 3 and 5 (Figure 6; hereafter referred to as "primary tunnels") were excavated in rounds of 1.0 m and supported by a steel-mesh reinforced shotcrete lining, which formed a closed ring able to undertake the overburden loads and reduce the surface disturbance. After excavation of the primary tunnels till the opposite diaphragm wall, reinforced concrete beams were constructed inside the primary tunnels. The beams were filling almost the entire tunnel cross section and served as pillars for supporting later the shotcrete shells of the secondary tunnels (2, 4 and 6 in Figure 6). During excavation of the latter, the shotcrete shells of the primary tunnels were partially demolished (Figure 7) and the shotcrete arch of the secondary tunnels was founded directly upon the prepared reinforced concrete pillars at pre-formed corbels. Figure 6 shows the individual concrete beams and the entire underpinning slab.

The design calculations showed that the microtunnels could be safely constructed even if the open pit outside the basilica had be excavated up to the foundation level of the slab -2, which is about 5 m lower than the level of the reinforced concrete slab. However, it was

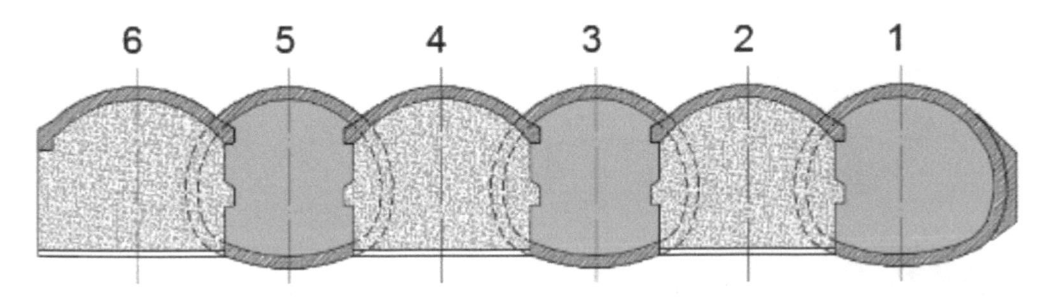

Figure 6. Intersecting microtunnels.

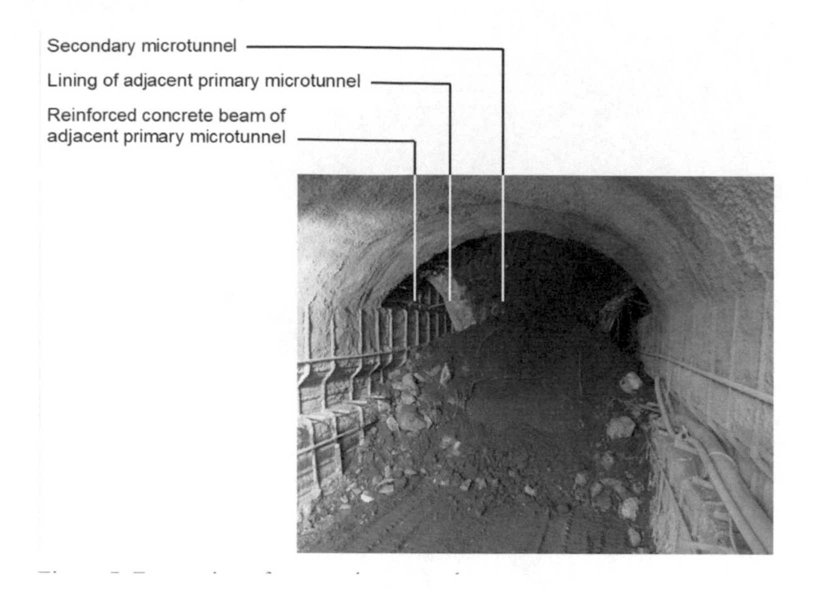

Figure 7. Excavation of a secondary tunnel.

decided, first to protect the floor of basilica by backfilling and then to excavate the microtunnels ahead of the excavation of the station underneath the slab -1.

Once the basilica was backfilled the microtunnels were constructed and filled with concrete and then the excavation underneath slab -1 was commenced up to slab -2. From this level the transverse secant piles were demolished and the excavation below the microtunnels was carried out. Upon completion of this excavation the slab -2 at the level -7.04 m was poured in one phase for the entire length of the station (Figure 8). Then, the excavation and construction of the station was carried out following the initially envisaged method up to the raft slab level.

5.3 *Main challenges and mitigation measures*

In the design phase, particular attention was paid to the potential hazards of, (i), face or crown instabilities during microtunnelling and, (ii), inadmissible settlement of the diaphragm wall.

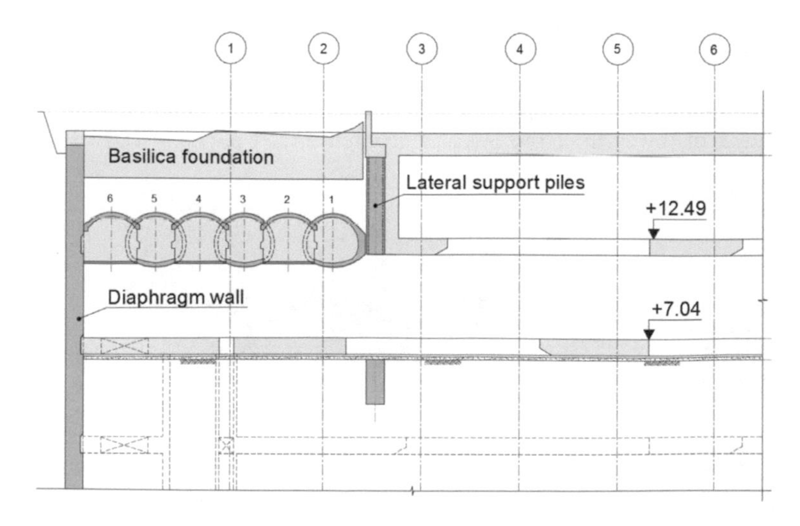

Figure 8. Construction stage: excavation under the microtunnels and casting the slab -2 (level +7.04).

5.3.1 Face or crown instability

In spite of the small cross section of the microtunnels, the experiences from previously constructed stations (where some monuments were founded on poor-quality manmade deposits) along with the presence of the water table at the level of the church raised concerns about the stability of the face and of the crown.

In order to investigate the conditions between the microtunnels and the basilica foundation, timely identify loose material and, if required, drain the soil, six horizontal exploratory boreholes were executed, each located at the crown of a microtunnel and being drilled along the entire microtunnel length of about 21 m. Furthermore, each microtunnel was excavated under the protection of an umbrella of spilling rods, providing continuous crown support during face advance. During the drilling of each probe the return material was checked and the response of the drilling bit was evaluated. The drilling diameter was 132 mm and supplied with casing to ensure borehole stability. At the end of each drilling a steel bar diameter 32 mm was inserted and grouted. From this process no weak soil strata was found.

5.3.2 Inadmissible diaphragm wall settlement

The self-weight of the station, the overburden weight above the roof slab, the live loads (traffic etc.) and the vertical component of the earth pressures are being transferred directly to the soil via the vertical diaphragm walls. The last excavation phase is the most critical one, because the embedded length becomes minimum while the loads from the superstructure and the earth pressure are maximum. The embedded length of the diaphragm wall was defined at the early stage of the design (before the discovery of the basilica) based on empirical values (DIN and EAP) for the tip and shaft resistances. Based on this analysis the diaphragm walls were constructed. With the new concept the diaphragm walls had to undertake significantly higher loads due to the overburden above the microtunnels plus the increased thickness (~2.5 m) of the underpinning slab. These new load conditions raised concerns and the verification against vertical stability was questionable.

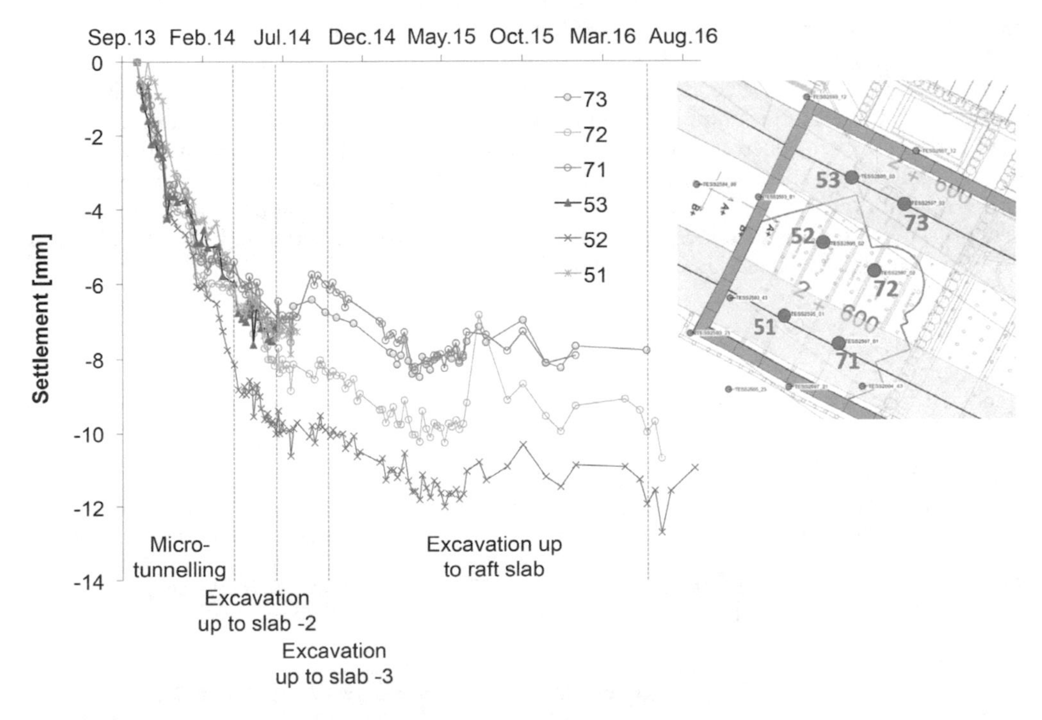

Figure 9. Surface settlements above the backfilled basilica.

Considering that the empirical values from several codes and national recommendations are in general on the safe side, under the new conditions more realistic strength values should be assessed. At the time of station re-design, the results of micro-pile suitability load tests were available. The latter had been performed at the vicinity of the station for the assessment of the ultimate resistance of the micro-piles, which would be used to the anchor the raft slab of the Sintrivani crossover (a deep shaft at a distance of about 350 m from the station). From the back analysis of the micro-pile test results significantly higher values for the diaphragm wall shaft resistance were deducted after applying appropriate correction factors. These new interface shear strength values were used and the bearing capacity of the diaphragm walls under the increased vertical loading was verified with sufficient safety factor. Excavation of the station was carried out safely and, as shown by the intensified monitoring programme, without any settlements of the diaphragm walls. Furthermore, additional micropile suitability tests, which were carried out at the level of the present station raft slab, confirmed the assumption about the shaft resistance.

Figure 9 shows the measured surface settlements above the backfilled basilica. It can be seen that the major part of the settlements, 6 - 8 mm, occurred during the construction of the micro-tunnels. The subsequent construction stages caused additional settlements of about 4 mm.

6 CONCLUSIONS

The Thessaloniki metro project is complex due the poor ground conditions along part of the route, the artesian pressure, the in situ hydraulic head gradient and seepage flow, the seismicity of the region and the immediate vicinity of buildings and monuments, which – besides being sensitive to induced settlements – impose severe constraints to the layout of the stations. The Sintrivani station is a typical case of the first five stations (four of them situated inside the walls of the old city) of the Thessaloniki Metro where the archaeological layer extends up to a depth of about 11 m below the surface. The survey revealed an important Byzantine church, which the Authorities decided to preserve due to its historical importance. The important archeological findings were discovered in the beginning of 2011 and put the construction of the station on hold for several months until the Authorities and the Client took the final decisions. The station was re-designed and excavated successfully underneath the basilica in the period 2012–2017, with the smallest possible impact on the initially planned station layout itself.

The design made use of the already constructed diaphragm walls and a minimal solution was developed to underpin the finding. The intersecting microtunnels were excavated, supported and filled with reinforced concrete through the limited space provided by the access northwest shaft.

The small access shaft affected the logistics with impact on the progress. The construction of the microtunnels started in September 2013 and took about 7 months including the construction of the reinforced concrete beams. On average the excavation and support of each of the 21 m long microtunnels lasted about 40 days.

REFERENCES

Brandl, H. 1989. Underpinning. *12th Int Conf on Soil Mech and Foundation Engineering*, Rio de Janeiro, 2227–2258

Hessler, N. 2012. Durchmesserlinie Zürich – Unterquerung Südtrakt «Bergmännische Deckelbauweise» auf dem Prüfstand. *Swiss Tunnel Congress*, Luzern.

Makropoulou, D. 2014. The findings of Byzantine Thessaloniki revealed during the works on the construction of the metropolitan railway of the city. *34th Symp. on Byzantine and Post-Byzantine Archaeology and Art, Athens.*

Vassilakopoulou, G., Rizos, D. & Vrettos, Ch., 2010. Thessaloniki Metro: Geotechnical and structural design of stations constructed using the top-down Method (in Greek). *Proc. 6th Hellenic Conference on Geotechnical & Geoenvironmental Engineering, Volos.*

Vrettos, Ch., Vassilakopoulou, G., & Rizos, D. 2013. Design and execution of special foundation works for the deep excavations of the Thessaloniki Metro. *Geomechanics and Tunnelling. 6–2013*

Tunnels and Underground Cities: Engineering and Innovation meet Archaeology, Architecture and Art, Volume 1: Archeology, Architecture and Art in underground construction – Peila, Viggiani & Celestino (Eds)

Rome and its stratification

G. Romagnoli
Sapienza Università di Roma, Italy

ABSTRACT: The research and the drawings of this intervention are part of a cycle of experiments series, which are feasible, though unusual, and look, making use of architectural instruments, for some possible solutions to important "urban paradoxes" in Rome. In particular, I refer to three of them: 1. the city backwardness in applying communication methods between its actual and historical forms aimed to design new urban layouts; 2. a renewed acknowledgement of the Tiber role in the city after the detachment due to the high walls built by Canevari on the river banks at the end of XIX century; 3. the lack of control of the city suburbs development and the progressive flacking of the urban form. The three parts of this research deal with them.

1 INTRODUCTION

The relationship between architecture, art and archaeology is an interesting, fascinating but sometimes unstable conceptual territory. It is rather difficult to trace clean borders between these human activities but, beside to this perhaps fruitless question, the real novelty nowadays is a reversal in the relationship between architecture and art. In some cases we can witness real intrusions of art into architecture fields.

As examples of works of art created in our times with architectural instruments, I just refer to two well known ones: 1. the repetitive and modular languages in the "land art" installations of Christo and Jean Claude, where they recall images of endless infrastructures deployed in the most suggestive world landscapes; 2. "The Seven Heavenly Palaces", closed off towers made of reinforced concrete, permanently exhibited at the Pirelli Bicocca Hangar in Milan. In brief, art is not a substitute for architecture.

Many archaeological finds of architecture, by now deprived of their previous purposes, have become art objects: this applies either to the different approaches of XVIII century "vedutismo" (landscape painting) by Piranesi and van Wittel or to the photos of industrial and infrastructural archaeologies made by Berndt and Hilla Becher, now exhibited at the New York Moma. Also in the past, Michelangelo, architect and sculptor, orderly conjugated architecture and art: in particular architecture and sculpture in the Sagrestia Nuova of San Lorenzo Basilica in Florence and modern and ancient architecture in Santa Maria degli Angeli e dei Martiri in Rome, where the relics of the Diocletian thermal system were changed into a christian church in order to give them back to the city vital cycle.

Nowadays leanings come up to substitute architectural with art works and this attitude change tangencies into cuttings between these two human activities. In many cases this practice produces magnificent works, which are also highly evocative, but sometimes they may reveal themselves inadequate from a technical or functional point of view, also as far as security is concerned. At this regard, I want to refer to Anish Kapoor project of the entrance to the underground station of Monte Sant'Angelo which is under construction in Naples. At symbolic and aesthetic levels. I think that this may well be the best work proposed for that purpose but I wonder how that project shape can protect

Figure 1. Christo, Jeanne Claude, *Runnig fence*, California, 1972.

the entrance to the underground areas from the draughts because it fosters instead of cutting them.

Architecture is "spectacular" when it overcomes one's expectations taking advantage of some requisites: the use of abandoned building artefacts; self-maintenance capability; access facilities; interactions with the existing housing; multitasking paradigms (integrating fast and slow mobility); the return of archaeological relics to the vital cycle of the cities; the production and use of clean energies).

The research and the drawings of my intervention are part of a cycle of experiments series, which are feasible though unusual, and look, making use of architectural instruments, for some possible solutions to important "urban paradoxes" in Rome. In particular I refer to three of them: 1. the city backwardness in applying communication methods between its actual and historical forms aimed to design new urban layouts; 2. a renewed acknowledgement of the Tiber role in the city after the detachment due to the high walls built by Canevari on the river banks at the end of XIX century; 3. the lack of control of the city suburbs development and the progressive flacking of the urban form.

The Roman archeologic heritage has been, and still is, one of the main causes which make infrastructural and building design extremely complex in the city. In order to face these questions, three analyses are proposed here: *Roma Ipogea, Roma Tiberina* and *Roma Aerea.* They are not real projects, but rather spatial diagrams, zero degree architectures, which pinpoint the potentiality of my research outcomes.

2 ROMA IPOGEA

This research as to do with the first urban paradox: the city backwardness in applying communication methods between its actual and historical forms aimed to design new urban layouts. Roma Ipogea suggests a dialogue between the actual city with its archaeological layers through a third deeper pedestrian route which goes from the Spanish steps underground station to a hypothetical one in the centre of Venice square (Figure 2).

Therefore we have three levels: 1. the actual city; 2. the system of new entrances and spaces sited on the archaeological level, seven meters below the actual urban one; 3. a pedestrian channel, one and a half kilometer long, thirty six meters deep, rich of services. This grid pattern, which is below the archaeological level, may be expanded.

Figure 2. G. Romagnoli, *Roma Ipogea, Piazza Venezia*, 2010.

Roma Ipogea identifies, through these three overlapping levels, different areas which promote concealed parts of archaeological relics, systems to access them either by feet or visually and other empty potentials which let air and natural light into the deepest level gaining a spatial continuity either horizontal or vertical.

The urban level shows only bare parapets aimed to ring holes which have the vital roles of windows for the underground levels. This design involves digging out, one of the most radical architectural practices. The light is most valuable in the cavity where its relationship with the bottom.

The layout in its entirety shows a system of empty spaces which are apparently inconsistent. Their forms and dimensions are associated to of the actual city to the *Forma Urbis Romae*, the map of the ancient city drawn up by Rodolfo Lanciani between 1893 and 1901. Out of the many layers of Rome, those are the meshes, the scans which have been considered basic in this process. In short, it is a sort of a ghost project, or of artificial architecture, as Franco Purini would call it, referred to a city found again, as those buried from time but still alive, a metaphor of the resurrection of the ancient Rome, which comes back through a slow infrastructure to be traversed by feet. The realisation of this "slow" infrastructure would require anyway important financial resources.

The third level is less restricted *vis à vis* the other two which are bound to the pre-existing layers of the city. Here the model of the "new main street" is adopted; it can be seen in the modulation of the cross-sectional route design (Figure 3). It consists of three drawn close squared modules, twelve meters on the side, which make up a comprehensive section thirty six meters wide and twelve meters high. The pedestrian flow is provided in the central module while the two lateral ones host the services structures which are inhabited as in a new village. Here a new underground part of the city is conceived where a plurality of programs are coordinated with a multitasking scope. These spaces do neither follow the "no places" model nor that of the alienating sales centres, they are rather shaped up to be meeting or educational places.

This underground route has five points of convergence: 1. A study centre at the Spanish steps; 2. An artisans' market which recalls the city craftsmen streets; 3. An audiovisual centre under the Gassman theatre in via delle Vergini; 4. A museum of ancient art under the empty space of the most Saints Apostles; 5. A large glass display case crossed by the load-bearing escalators which lead to the Venice Square underground station.

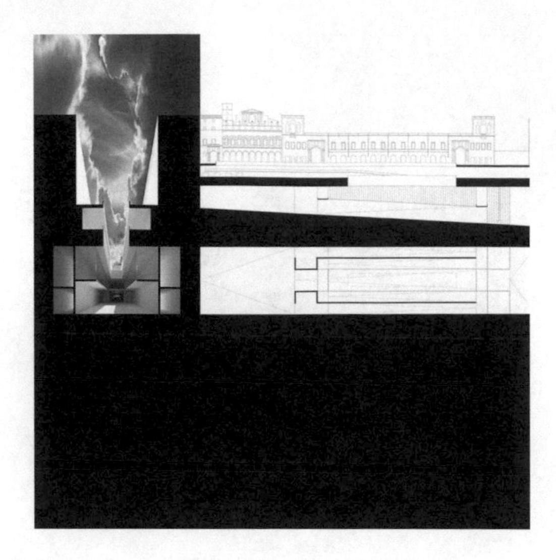

Figure 3. G. Romagnoli, *Roma Ipogea, Museo SS. Apostoli*, 2010.

The analysis of the city, in its axes, crossroads and forms shows how this project proposal is linked to its past, to its catacombs. Its two main scopes are to live a furrow and to find the light. So the holes function as windows of a long edifice which lies horizontally and looks at the sky from the bottom. These elements belong to the genetic code of Rome, from the mythical furrow cut by Romulus to the emblematic *oeil-de-boeuf* represented by the Pantheon, the icon of the city.

In 2010, *Roma Ipogea* has opened again, after ten years, the question of an underground station at Venice Square as a fundamental focus (Figure 4). Just under thc central flowerbed a plurality of archaeological relics had been found: the statues of two magistrates, a II century a.d. street of *lapis gabinus*, a glassmaker of the XVII century and more, under the secular pines on the side of *Santa Maria di Loreto*, the *Adrianeum* flight of steps.

Figure 4. G. Romagnoli, *Roma Ipogea, Piazza Venezia*, 2010.

Despite a fund of 900 millions of euros and the works 40 meters deep, already begun in 2000, the Monument and Fine Arts Office decided to cover the dig. The underground station of the line C was provided in the green open space under the Trajan column. Any possible ambition of reopening the question, Venice Square has been covered with reinforced concrete and therefore the traffic may freely flow again, without any stumbling block, in the "city of the ostrichs".

The underground station of *Roma Ipogea* is sited in this system centre just under the green oval of the square where the actual city is to give impulse and trust in rewriting its stratigraphy though going down. Meanwhile the settlement of the lateral lay-by on the right, which had been deprived of its secular pins and where the *Adrianus* pre-existences are located, risks to remain unsolved. One purpose of this project is to react to the inertia of a city which, differently with others, does not take advantage its underground layers in order to localise infrastructures necessary to its development.

3 ROMA TIBERINA

At the origin, the Tiber played a determinant role on the city location but the recurring floods have often turned it into a calamity.

Roma Tiberina deals with the second urban paradox: a renewed acknowledgement of the Tiber role in the city after the detachment due to the high walls built by Canevari on the river banks between 1876 and 1901. They were decided by the Savoia, the royal family, after the overflow of December 28 1870, and have built a barrier between the river and the city. They have made almost invisible the Tiber inside the city, a "built in river", as Ludovico Quaroni named it, canceling almost completely the interaction systems between the people and the river, the vital rhythms of the millstones, the ports, the river banks. The building of the high walls, beside a visual detachment, has brought to demolition of portions of the city, including the infrastructures system, either big or small, which revolved around the Tiber and through which the city lived its river.

Among them, we recall the ports of Ripetta and Ripa Grande, the Doganella building, the millstones, the descents to the river. Further more the Altoviti Palace, a XVI century splendid residence, and many other palaces directly grounded into the water were destroyed. The Roesler Franz watercolours offer a vision of modern Rome as a large boat.

Therefore this research has been finalised to indicate a possible role of the river within a logics of its new connection with the urban system on several levels, in continuity of thought and urban city planning with *Roma Ipogea*. The design wants to reject some common places, as the archaeological conservation issue which block the city driving force motion, and to favour its natural development making use of resources still inexplicably neglected (the river, its banks, the high walls, the hypogeum river). This project is strategic because it updates and order many potentialities (the river banks regularisation for pedestrian access on both sides of the river, opening of new structures along the river, use of hypogeum spaces below the riverfronts, the realisation of a fast transport infrastructure along the river banks, new bridges in the urban tract, the full navigability of the river linked to the water jump removal beside the Tiberine isle) (Figure 5).

The river, as a sliding street, becomes a control and transport instrument. So, in addition to a natural element, the Tiber changes into a development axis.

The research on the river reveals the symbiotic evolution of the river and the city. The Tiber is analysed from its myth to the *ph* of its waters. The intersection of several contexts to which the river is linked represents the environmental complexity of an eventual city planning intervention.

The study insists on the river portion between the Margherita and Sublicio bridges, the oldest segment of the historical city. Along this tract of the river, some cross sections have been realised in order to verify the possible localization of fast transport infrastructures (Figure 6). In the fourteen meters between the actual buildings foundations and the river

Figure 5. G. Romagnoli, *Roma Tiberina, Rampa Tiberina*, 2015.

Figure 6. G. Romagnoli, *Roma Tiberina, Cross sections*, 2015.

banks, a gallery with a single track, six meters wide, can house an underground line (D) which connect the whole city touching it at its crossroads and having its stations beside the bridges. To the North, it would intersect the line A (at the Flaminio station) by the Margherita bridge. To the South, it would intersect the underground line B (at the Piramide station) by the Sublicio bridge.

The Umberto port, sited under the namesake bridge, would take advantage of the open space between the Tor di Nona and Marzio riverfronts. This new square would be a dock which recalls the Ripetta port buried under the filling earth of the canevarian project. The urban level is connected to the river either through the access to the hypogeum level in front of the Napoleonic Museum or through the symmetric open space where a kiosk is actually sited. Inside the bridge foundation, under its slope, a large flight of steps finds its room leaving space to a passage on the river bank.

Three bridge-squares lean against the high canevarian walls. Each of them, one hundred meters on the side, hosts gardens which compensate the city centre with green areas. The first garden square surrounds and embrace the Sant'Angelo bridge, connecting it to the green areas of the namesake Castle today neglected. The second green bridge places itself between the Villa Farnesina and the Farnese Palace and finds a connection with to the Botanical Gardens. The third bridge places itself in the position of the ancient Sublicio bridge, as reported in the Nolli's maps, and connects the Santa Francesca Romana's garden, beside Santa Maria in Cappella, with the Aventino steep hill, beside the slope of Rocca Savella and the gardens of the Olitorio Forum.

The last planning sight concerns the Tiberine Ramp, which connects the city with the river banks through an architectural *promenade* earned inside the thickness of the high canevarian walls and give access to a museum dedicated to the Tiber. This museum places itself in front of the Tiberine Isle on the side of Fabricius bridge, where Rodolfo Lanciani reports the relics of San Abbaciro ad Elephantum, a V century a.d. church, which was demolished in the XVIII century.

This complex is located around the church in the hypogeum section inside the canevarian walls; it conjugates the historical, modern and contemporary memories of the river.

An important information of this research on the river regards the Paleotevere, a large flow aquifer known by the geologists since a long time. It makes up the Pleistocene riverbed and is placed between two waterproof ground layers from thirty to sixty meters deep below the actual river level (Figure 7). This underground aquifer flow rate is both minimal and unpolluted. One interesting aspect is that it can produce geothermic energy. The waters of the ancient Tiber reach 20° Celsius, on average, and their temperature may be used to tap heat energy. According to some seismologists – Franco Barberi's equipe of the National Institute of Geophysics and Volcanology at Roma Tre University – say that "it would be easy and convenient to extract the geothermic liquids, direct them to the apartment buildings and, after tapping the energy through heat pumps, inject them again in the aquifer to avoid its depletion" (a.t.). Their research was anyhow interrupted in 2010, the motive being likely to be found in the economic interests linked to energy provisions which are alternative to the fossil type.

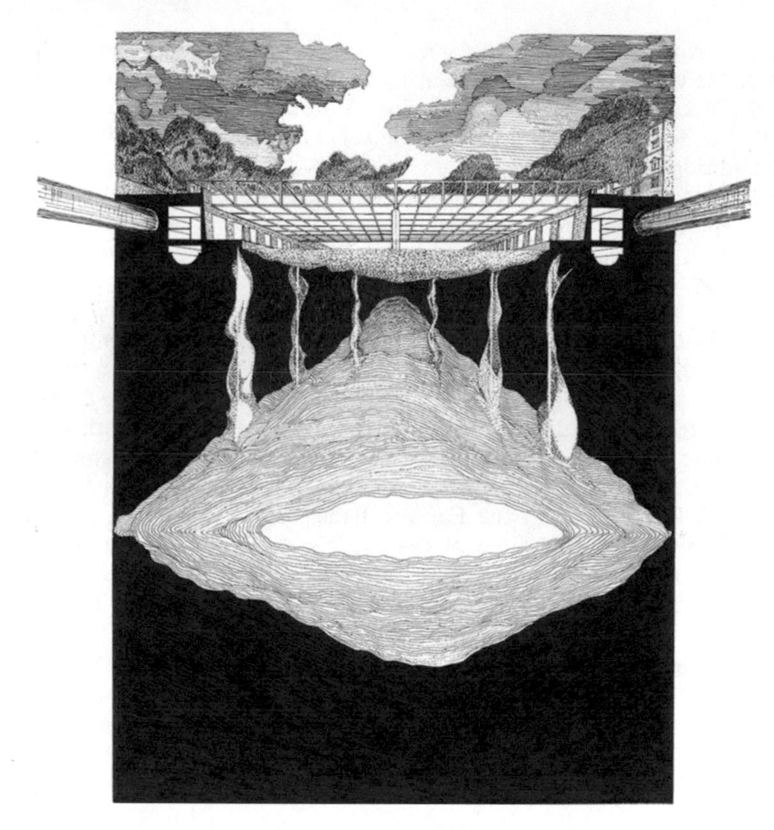

Figure 7. G. Romagnoli, *Roma Tiberina, Ponte Farnese*, 2015.

4 ROMA AEREA

This part of the research deals with the third urban paradox: the lack of control of the city suburbs development and the progressive flacking of the urban form. The aim is to contribute to a renovation process of the urban system with territorial strategies which pay adequate attention to the housing spaces, the environment, the infrastructures.

Looking at Rome from the air, it is impossible to identify in the evolutive cell of the urban system a form related to an effective city planning. The only control elements which still, though hardly, seem to withstand its proliferation towards the peripheries are the radial grids designed by the old consular streets which depart from the city centre and cross the Great Ring Road (GRR), the last stronghold attributable to a contemporary form of the city, an instrument of urban identity today made weaker and weaker by urban speculation and illegal development.

The question is whether it is still actual and worth to argue about the urban form of Rome. Is the GRR just and only an infrastructure relegated to a role of mobility and clearing or it may be considered the heir of the custom belt previously represented by the Squared Rome, afterwards by the Servian Walls and finally by the Aurelian Walls?

Among the endemic diseases of Rome, we find a high degree of anarchy which is responsible for the absence of an adequate city planning, except for the city walls mentioned above, which have mainly had a restraint function.

The first table of Roma Aerea represents a layout of the city, inside the GRR, as it appears at night (Figure 8). It is a strong image in the contrast between full and empty spaces and anti-iconic at the same time. As in a star dust, one cannot trace the city celebrated monuments which are confounded with the other buildings. The table returns the image of a fragmented city, sublimated by the distance, where the relationship between in and out is dumb.

Figure 8. G. Romagnoli, *Roma Aerea*, 2017.

The GRR, like an ancient wall, marks a complex system, where an ideal layout of a circumference bends itself following the morphology of the territory. Along the Aurelian Walls a system of towers and doors strategically articulated a "rhythm" around the city.

Today it is possible to indicate other city centres around the GRR as strongholds aimed to reinforce the identity of the peripherical systems where they insist.

Over this border, between the Appian and the Tuscolana ways a new door places itself, a linear edifice located between the XIX century Capannelle stadium and the Ciampino airport landing strip. The design at the bottom of Figure 8 shows an axonometric projection which joins the three elements. The aim of this projection is to give a unitary character to a sequence of disarticulated spaces, by creating a point of convergence which offers a new centrality to the periphery. The three areas should go on and maintain the roles previously assigned to them: the airport, the hippodrome, a place for big events. The unifying sign tends to enhance and give value to these structures to which others can be added either commercial or cultural. The leap in scale associated to this operation, capable to ideally recall the great Roman aqueducts which design this tract between the city and the countryside, raise a new discussion on the GRR role. A new relationship may be created between the inside and the outside of the urban form together with the possible future localisation of further poles along its perimeter.

5 CONCLUSIONS

These three studies aim to give Rome a pushable to move it from the contemplation of its ancient ruins to will and knowledge such that it may recover a new flame through the enhancement of the primordial elements: water, earth and air.

REFERENCES

AA. VV. 2006. Il Tevere a Roma, Portolano. Milano a cura dell'Autorità di bacino del fiume Tevere. Edizioni Ambiente

AA. VV. 1985. Tibre-Seine: deux villes deux fleuves. Le Tibre. Roma. Carte Segrete editore

AA. VV. 2008. Uneternal City, Urbanism beyond Rome. Venezia. Marsili

D'Onofrio C. 1968. Il Tevere e Roma. Roma. Ugo Bozzi

Ferlenga A., Biraghi M., Albrecht B. 2012. L'architettura del mondo. Infrastrutture, mobilità, nuovi paesaggi. Bologna. Compositori

Funiciello R. 2007. I caratteri geologici dell'area romana, in, Il territorio tra la via Salaria, l'Aniene, il Tevere e la via "Salaria vetus", (a cura di) Cristiana Cupitò, Quaderni della Carta dell'Agro Romano nr. 1. Roma. L'Erma di Bretschneider

Lanciani R. 1989. Forma Urbis Romae. Roma. Quasar

Lugli P. M. 2006. L'agro romano e l'altera forma di Roma antica. Roma. Gangemi

Ryckwert J. 2002. L'idea di città, Antropologia della forma urbana nel mondo antico. Il ramo dell'oro, Adelphi

Segarra Lagunes M.M. 2004. Il Tevere e Roma. Storia di una simbiosi. Roma. Gangemi

Tunnels and Underground Cities: Engineering and Innovation meet Archaeology, Architecture and Art, Volume 1: Archeology, Architecture and Art in underground construction – Peila, Viggiani & Celestino (Eds)
© 2020 Taylor & Francis Group, London, ISBN 978-0-367-46574-2

Line C in Rome: San Giovanni, the first archaeological station

E. Romani, M. D'Angelo & V. Foti
Metro C, Rome, Italy

ABSTRACT: The construction of Line C San Giovanni station in Rome required the adoption of building techniques suitable for responding to the numerous problems concerning geology, archaeology and the surrounding buildings. The procedures for building this station were heavily constrained by what emerged unexpectedly at the start of the work: the archaeological findings down to a depth of 18 m from ground level. The project faced logistical challenges, such as the underpassing of the existing Line A San Giovanni Station with two tunnels excavated by conventional mining using artificial ground freezing. The archaeological findings required an overall revision of the internal design of the station: at the end of the archaeological excavation, the Archaeological Ministerial Office required the contractor to change the interior design to safeguard these findings. *Metro C* designed a new architectural layout and San Giovanni station became the first archaeological station in Italy.

1 INTRODUCTION

Line C is the third line of the Rome underground. Once completed, it will run under the city from the South-eastern to the North-western area, for a total length of 25.6 km and 30 stations. *Metro C* S.c.p.A—the General Contractor led by Astaldi—manages the construction of Line C in all its phases: design, archaeological surveys, tunnel boring, excavation and construction of stations, building of trains and start-up.

At present, 22 stations and 19 km of line are open to the public.

San Giovanni station required the adoption of particular building techniques suitable for responding to the problems of geological/geotechnical, archaeological, and territorial nature, because of the high density of buildings and underground utilities. The station is located in the urban environment of the Appio - Latino neighbourhood, and extends along Via La Spezia next to the existing Line A San Giovanni station.

It is 140 m long and 22 m wide and reaches a depth of 35 m from ground level. The structure, with its 7 floors, is adjacent to the existing Line A station, allowing passenger interchange between Line A and Line C.

The perimeter diaphragm walls have an excavation length of about 56 m and reach the Pliocene clays in which they fit at the base for about 12 m.

2 GROUND CONDITION

The geological/geotechnical units encountered while excavating range from the volcanic units of the Eastern outskirts of the city to the fluvial or marine units of the historic centre of Rome (the Tiber's ancient riverbed).

Starting from ground level, the first stratum consists of made ground [R], with a thickness of 13 – 16 metres. It is a heterogeneous material with a sandy-silty matrix, mainly of pyroclastic nature. These materials offer fair resistance to effective stresses as well as modest cohesion.

This is followed by a stratum of recent alluvial deposits [LSO] that have filled deep cuts of often variable thicknesses, as thick as 17 m at San Giovanni station. In terms of granulometry, it consists mainly of low- medium stiff blackish silty sands and clayey silt. The succession of the units composing the Pleistocene deposits, proceeding from top to bottom, presents an initial unit [ST] consisting of an alternation of white or yellow silty sand and clayey and sandy silt [STa], with carbonate concretions and localised levels of travertine concretions, with a high degree of thickening and of medium to high density, followed by a very stiff unit consisting of clayey silt and silty clay [AR] of grey or yellowish colour due to oxidation.

The lower unit of Pleistocene sediments consists of very dense, medium to coarse sand with gravel [SG], with its roof localised at + 6 ÷ 11 m below sea level.

This stratum of sand with gravel lies directly upon the Pliocene formation of very stiff grey-light blue silty clay and clayey silt [APL], whose top rests at the depth of 44 metres from ground level. In the first 5/6 metres of the unit top, the stratum consists of some levels of very dense fine sand, alternating with compact clayey silt of different thicknesses, no more than a decimetre thick. Proceding downward, the facies type becomes purely silty/clayey and highly stiff.

Site investigations show a relatively complex piezometric regime in accordance with the stratigraphic situation, in which erosion phenomena have partially or totally removed the sandy and clayey strata of the Pleistocene formation. The buffering effect into this formation does not always appear effective.

The measurements obtained by Casagrande piezometers or open wells, installed at different depths into various formations, or by drainage tests, generally demonstrate that the alluvial deposits and the made ground are the site of an unconfined aquifer with piezometric level at the absolute depth of 25.00 ÷ 28.15 m below sea level.

The measurements indicate the westward flow of the aquifer. The deep unit, consisting of sand and gravel, is the site of an artesian aquifer, supported by basal Pliocene clays and plugged from the roof by the clayey levels of the Pleistocene formations, whose piezometric level is at a maximum depth of +17.20 ÷ 17.66 m below sea level.

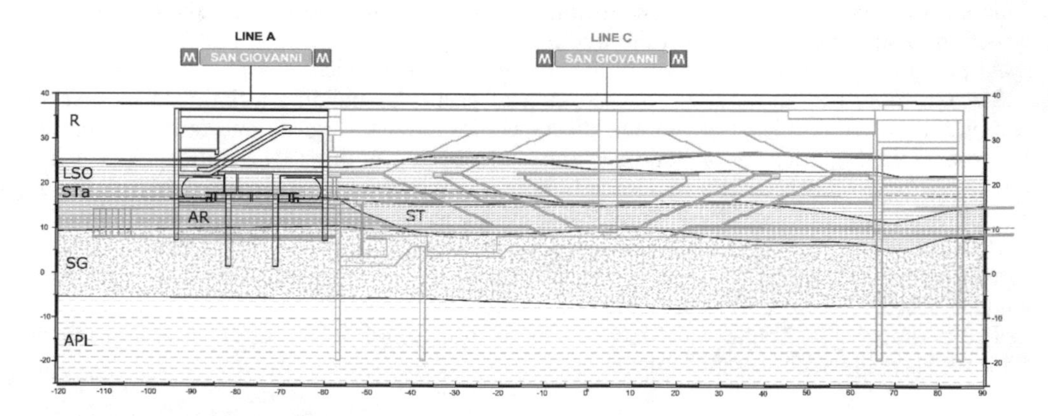

Figure 1. Geotechnical profile.

3 PRE-EXISITING BUILDINGS

The presence of a considerable made ground stratum makes the buildings adjacent to the station particularly sensitive during the excavation phase. These buildings were constructed in the early twentieth century. Their bearing structures are in masonry, and their shallow foundations are not sunk into the deeper and more compact strata.

Moreover, approximately in the mid 1950s, some of the buildings surrounding the station were elevated with additional reinforced concrete structures, without upgrading the existing foundations.

These buildings, with their masonry bearing structure, steel-beamed floors along with hollow flooring blocks, Roman-style stairways, and presumably shaft foundations, present more or less widespread cracking phenomena.

The station diaphragm walls, far deeper than the bottom level of the buildings, are positioned at a distance ranging from 1.80 to 3.00 m from the façade of the buildings.

A structural and geotechnical foundation reinforcement with 300 micro-piles was needed for the building named L3–Carducci school—because of major cracking phenomena, due to the settling of the bottom level, to the interaction with the station excavation and with the line tunnels running 100 m along Via La Spezia.

Among the pre-existing structures, mention must be made of the existing Line A San Giovanni station. This structure, placed in the middle of Piazzale Appio, stands between 80 cm thick diaphragm walls. Its rectangular plan is 75 × 30 m, and lies at a depth of 22 m below ground level. The interior bearing structures are made of reinforced concrete beams, steel columns, and brick/cement slabs connected to the perimeter partitions by pockets and/or grouting. The columns are based upon piles Ø2000 in diameter, sunk into the stratum of sands and gravels. Right beneath the entrance level, the existing station, designed in the 1970s, has always had a floor completely dedicated to the future passage of Line C.

The existing station was the subject of a highly in-depth study, to reveal the geometry of its structures and the reinforcement of the main structural elements, as well as the length and position of the foundation piles. The length of these elements was assessed by echometric tests performed from the under-platform area, while the topographical position was surveyed by partially demolishing the upper concrete cover of the foundation slab, and reconstructing the pile surface starting from the rods in the reinforcement cage.

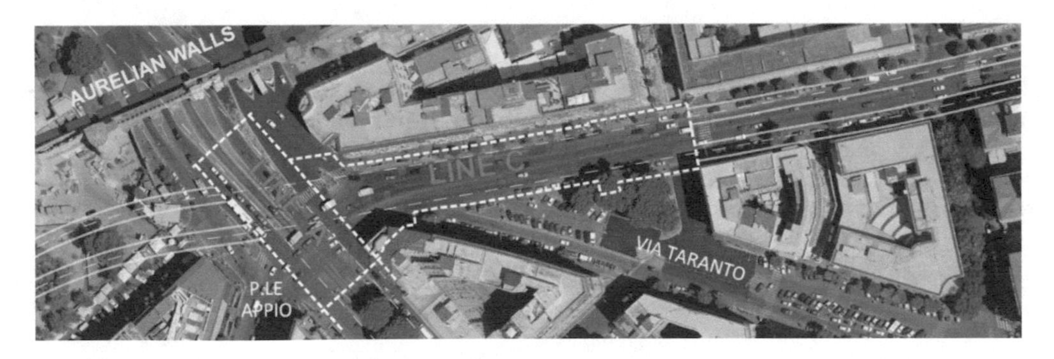

Figure 2. San Giovanni station worksite.

4 ARCHAEOLOGICAL SURVEY AND DESIGN CHOICES

The design choices in the construction of this station were heavily constrained not only by the pre-existing structures and the geological/geotechnical landscape discussed above, but also by what emerged during the archaeological surveys performed once the worksite had begun: unexpected archaeological strata distributed in landfill down to depths of 14 to 18 m, with a water table at about 8 m below ground level.

Metro C started the archaeological survey in San Giovanni station job site in 2007 with the execution of about 500 bore holes and the results of these surveys were reported in a document named "*Prontuario*", written together with the archaeological Superintendence of the Municipality of Rome. This document illustrates the excavation procedures and the cataloguing of the findings, so as to ensure the respect of performance times but at the same time guarantee all the requirements of historical and scientific nature.

A 3D reconstruction of the result of the archaeological survey in drawn in the Figure 3.

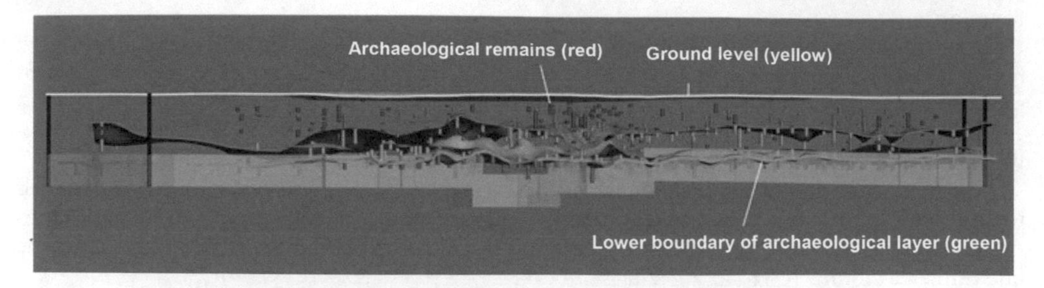

Figure 3. A 3D reconstruction of the result of the archaeological survey.

Following the results of the archaeological surveys, carried out before building the station diaphragm walls and executing the jet grouting bottom plug, the archaeological Superintendence of the Municipality of Rome imposed the following construction constraints:

1. impossibility of performing consolidations from ground level without prior archaeological excavation;
2. need to carry out all the excavations by archaeological method down to "virgin" ground.

These obligations required an overall review of the Line C design in the historic centre of the city of Rome, starting right from San Giovanni station, to safeguard the archaeologically sensitive strata.

5 DESIGN CHOICES

The archaeological findings compelled *Metro C* to change the design of the line that runs from San Giovanni to Malatesta and abandon the level left available in the Line A station.

In detail, a plano-altimetric variation of the Line was required starting from Lodi station, right before San Giovanni. The TBM tunnels were deepened beneath the archaeological strata, and traditionally excavated tunnels were introduced in order to pass beneath the existing station.

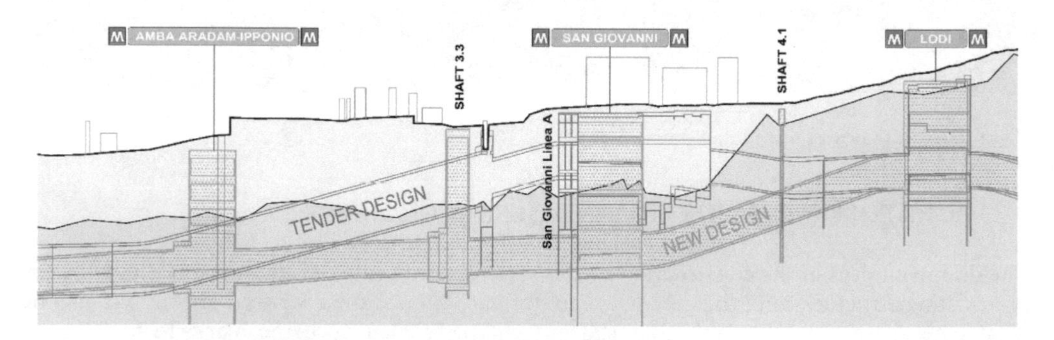

Figure 4. Comparison between the original design and the new design.

6 STATION EXCAVATION PHASES

The study of the new execution phases, heavily constrained by the archaeological findings, aimed at taking advantage of the benefits of the top-down method with the execution of descending slabs functioning as a strut to support the perimeter partitions. This led to defining the following succession in construction, which, for each slab, calls for:

1. excavation with archaeological method down to about 1.00 m beneath the slab inner face;
2. execution of a perimeter concrete edging resting upon pockets created in the diaphragms and hung on the covering slab using Dywidag bars;
3. deepening of the excavation with archaeological method for 2 additional metres;
4. execution of the slab with self-bearing prefabricated structures and subsequent casting.

These phases, by eliminating the excavation on the inner face of the slabs cast against earth which is typical of the "top-down" method, allowed the excavation to be deepened by horizontal levels, thus preserving the ancient structures.

For the self-supporting structures in the casting phase, mixed reticular precast beams were chosen, having spans of up to 20 m, along with self-supporting ribbed girders, with spans of up to 8 m and with a final slab thickness of 40 cm.

Figure 5. The construction on internal slab.

7 INTERACTION WITH THE EXISTING LINE A STATION

The interaction with the existing station of Line A entailed a series of design and construction choices.

The deepening of the excavation of the Line C station also generated problems of carrying capacity for the Line A diaphragm partitions. Given that there are no internal linings, the partitions transmit the vertical loads to the deep strata of the ground. To avoid this problem, it was deemed necessary to consolidate the stratum of sands and gravels at the foot of the diaphragm walls for the entire horizontal extent of the station by jet grouting. This expedient made it possible to build the Line C diaphragm walls adjacent to the Line A ones without altering the balance of the existing station.

To guarantee the transmission of the vertical loads over the long term, even after the execution of the tunnels passing beneath Line A, it became necessary to link the two structures with a reinforced concrete wall by underpinning, connected to the five Line C diaphragms and, by means of grouting, to the Line A partition.

Moreover, given the need to underpass existing station by means of a traditionally excavated tunnel abutting the foundation piles, it was necessary to assess the reduction of the lateral resistance of these elements. Since these were large-diameter piles (Ø2000) sunk into the extremely rigid stratum of sands and gravels, it was evaluated whether the carrying capacity at the end is preponderant with respect to the lateral carrying capacity. Moreover, the continuous 40 cm thick slab foundation guarantees a reduction in any displacement that may be generated.

Lastly, the two under-crossing tunnels cut the longitudinal diaphragm walls of the existing station, thus generating a reduced vertical carrying capacity for these diaphragms. To avoid the problem, a 40 m long wall beam was constructed, connected to the Line A diaphragm walls at the existing station platform level.

This structure, resting upon 3 plinths with preloaded micro piles, has the function of transferring all the vertical loads of the longitudinal alignment of the partitions to the deep strata of ground beneath the line tunnels.

As the traffic of Line A and the station operation could not be suspended while the tunnels were being built, a track support system, for both directions, was provided for at the platform level, built with a continuous metal deck with two 20 m long spans, resting on micro piles.

8 TUNNELS UNDERPASSING THE LINE A STATION

The project faced logistical challenges, such as the underpassing of the existing Line A San Giovanni Station with two tunnels, 40 m long each, excavated by conventional mining using a variety of consolidation techniques including chemical and cement injections and artificial ground freezing.

The design of the underpassing of San Giovanni Line A station was deeply affected by the need to maintain the train operation during the excavations of Line C tunnels.

The existing box-shaped station comprises 4 underground levels: it is made up of perimeter diaphragm walls and slabs in reinforced concrete. The tracks rest on a lightly reinforced 40 cm thick foundation slab, and between the two tracks are two alignments of steel pillars based upon piles 2 m in diameter and approximately 15 m deep from the slab level. The excavation profile of the Line C tunnels abuts on the base slab of the existing line, which is underpassed diagonally, and the tunnels interact with 4 piles of the existing station each. Each tunnel has an excavation section equal to approximately 55 m^2 and a length equal to about 35.5 m, with an interaxis of about 16 m.

The choice of conventional mining methods results from the presence of the slab and from the Ø2000 foundation piles of the existing Line A station which, as they are placed at a net interaxis of approximately 5 m, made it impossible to continue the mechanized TBM excavation (diameter 6.70 m). The choice of the freezing technology, on the other hand, results from the impossibility of carrying out consolidations from above, because of the existing station and, above all, the heterogeneous nature of the soils to be excavated. The latter, for about two thirds of the tunnel section, are sands and gravel, with a rather high permeability, at times even exceeding 10^{-4} m/s.

The technology of artificial ground freezing consists in freezing the water within a volume of soil, in accordance with a known geometry, thanks to heat exchangers in which low-temperature liquid circulates. This process extracts the heat from the ground and dissipates it to the outside. The particular complexity of the design geometries required the use of a mixed system employing both nitrogen and brine. Nitrogen was used in the initial phases of the freezing, to reduce the time required to form the ice wall since significant volumes of ground were treated, and brine in the maintenance phase, to control the growth of the ice wall in time, thus reducing the deformation induced on the existing slab of Line A.

Placed around each tunnel to be excavated were 36 freeze pipes, in which circulated a coolant, and 15 temperature probes, for the real-time continuous monitoring of the development of the ice wall. The distances between the freeze pipes varied from about 75–80 cm on the sidewalls to 3 m in correspondence with the invert, due to the piles; the average length was about 36–39 m. To respect the design's geometry and interaxes for their entire length, the drilling operations were guided using the TDDT (Trevi Directional Drilling Technology) system. The average design temperature was -10 °C and the ice wall made around each tunnel had a thickness of about 80 cm at the sidewalls and approximately 3 m on the invert.

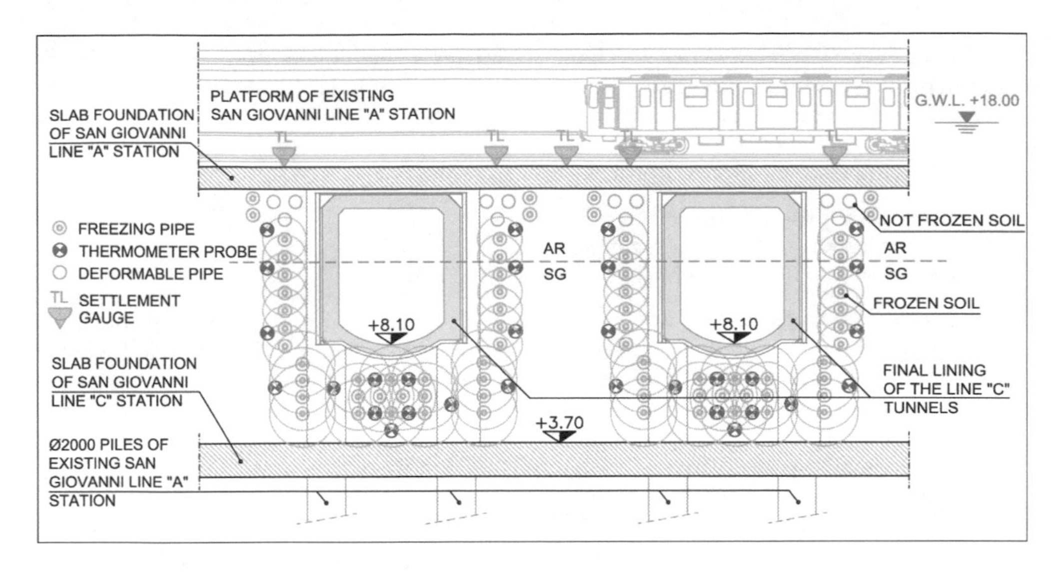

Figure 6. Cross section of the Line C tunnels underpassing the existing Line A station.

8.1 *The construction phase*

During the works of consolidation of the soil, unforeseen anomalous responses by the mass occurred, and were overcome by means of original technical and technological procedures.

The phase of guided horizontal drilling beneath the water table from the inside of the new San Giovanni station was marked by a number of operating difficulties due in part to the significant lack of homogeneity and the longitudinal and transversal discontinuity within the stratum of coarse gravel and sand, and in part to the presence of the magnetic field generated by the operating Line A. Due to these problems, directional drilling, magnetically guided with traditional asymmetrical tools using the TDDT (Trevi Directional Drilling Technology) system, proved ineffective. The guided drilling technique requires making a pilot hole through the advance by rotation, as well as the use of flexible rods and asymmetrical bits, while in the correction/deviation phases, the drilling is carried out by thrust only, without rotation, taking advantage of the asymmetry of the machine, appropriately oriented in the desired direction.

During the drilling operations, upward deviations of even 4 m were recorded after a 20 m advance, and therefore a number of holes were abandoned because they were incompatible with the design tolerance. New supplementary holes were drilled by trialling a variety of drilling techniques, with the recording of deviations for subsequent steps, if the technique allowed it, or alternatively afterwards.

These trials, after about 6 months of field testing, led to the choice of combining a Wassara-type water-powered hammer at the hole bottom with a system of direct drawing of final awaiting tubes. The use of special disposable linings for each drilling operation, with materials and joints oversized in comparison with what may be expected on the soils on site, and the choice of an innovative, solely percussion system, made it possible to stiffen the drilling system and limit the deviations of the holes.

This drilling methodology is certainly unconventional, and had never been used before in soils having geological characteristics similar to those in question. The previous experiences and various deeper examinations, gradually put into play by the specialized company Trevi S. p.A., based upon a gradual adaptation of techniques derived from other fields of application, made it possible to identify an original operating technique that was not in the original design forecasts, but that still allowed the initial design objective to be achieved in substance, albeit with adaptations of and supplements to the initially planned grout mask (Bertero et al., 2016).

In the final analysis, the drilling operations progressed quite irregularly. In correspondence with the worst sections, the vertical inter-axes (between the holes of the sidewalls) were increased to about 1.6 metres at the hole bottom, while the horizontal inter-axes (between the holes in the invert) were increased to about 3.7 metres at the hole bottom. The new freezing mask therefore involved an increase in the number of freeze holes by about 50%, and consequently an increase in the consumption of nitrogen and of the time needed to reach the ice wall thicknesses provided for in the design, due to the divergence between the freeze pipes.

The following freezing apparatus was installed:

- 17 temperature probes for each tunnel;
- 107 temperature sensors for each tunnel;
- 36 freeze pipes on the right tunnel and 39 on the left tunnel.

As already discussed, freezing was activated with nitrogen and maintained with brine. The brine freezing system consisted of 2 refrigeration plants, with a power of 250 thermal KW each.

The freezing activation and, similarly, the excavation of the two tunnels took place at different times. In both cases, the nitrogen activation lasted for about 40 days, and afterward the system was converted to brine. The nitrogen consumption for each tunnel was about 1500–1700 l/m^3, in line with the design forecasts, based on the interpretation of the data of a specific field trial.

Once the ice wall was formed, the diaphragm walls of both the new and the existing S. Giovanni stations were cut by a diamond wire saw. The tunnels were excavated with the temporary support of sprayed concrete and steel ribs at the roof and sides. This phase lasted for about 45 days, with an industrial production of 1.2 m/day. It is worth mentioning that in the excavation phase, the brine flow was modulated using the data surveyed with the thermometer probes, with the aim of governing the deformations induced on the Line A slab. Experience showed little effectiveness of the lightening holes made with deformable PVC pipes (Ø125) in correspondence with the sidewalls in order to allow the soil beneath the slab to expand.

Another particular aspect of the interaction with the existing structure is the foundation piles' interference with the track section of Line C. The problem was solved by re-profiling foundation piles, once the permanent lining of the tunnels was complete, by diamond wire, followed by horizontal cuts from top to bottom. In order to restore the continuity of the pile's reinforcement, steel bands welded to the pile's reinforcement, and a protection casing, were installed.

Figure 7. Freezing pipes and tunnel excavation.

8.2 *Monitoring results*

In light of the characteristics of the urban and structural environment affected by the works, and of the complexity of the project, the monitoring system played a key role for providing, as quickly and completely as possible, all the parameters needed to quickly reconstruct the phenomena in progress and their developments over space and time.

The size of the volume investigated and the quantity of parameters to be monitored were defined based on the design forecasts regarding the characteristics and the values of the interferences generated during the excavations for the works in question, as well as the performance of the consolidation technologies used (Ground Freezing), and so as to be able to effectively reconstruct the characteristics and development of the physical phenomena.

The variation in the volume of water in the passage from the liquid to the solid state exerted pressures on the existing station foundation slab. The movements of the slab were monitored using 18 settlement gauges placed along three alignments of the bottom slab of Line A, parallel to the rails. In general, the trends recorded on the three alignments were rather homogeneous and showed a higher deformability in the centre line of the slab, according to the restraint conditions on the slab.

The modelling and sizing of the interventions were correct, and the maximum movements recorded by the settlement gauges placed on the slab – on the order of 50 mm along the central alignment and 20 mm along the alignments near the diaphragm walls – were in line with the design forecasts.

It is interesting to examine the behaviour of the settlement gauges placed on the central alignment (alignment 2), where the two freezing phases of the two tunnels may be detected by the raising trends recorded. It is important to stress that in the nitrogen freezing phase, specifically in consideration of the monitoring data recorded, it was chosen to modulate the flows in the freeze pipes closest to the Line A slab, placed in the AR clays formation, in such a way as to reduce the effects on the Line A slab.

The passage to maintenance with brine and the completion of the excavation of the tunnels on both paths showed a substantial stabilization of the monitoring data.

The measurements of the movement of the diaphragm walls are negligible; while on the pillars, values on the order of 17 mm were recorded on the level above the Line A platform (third underground level) and of 10–14 mm on the level above (second underground level).

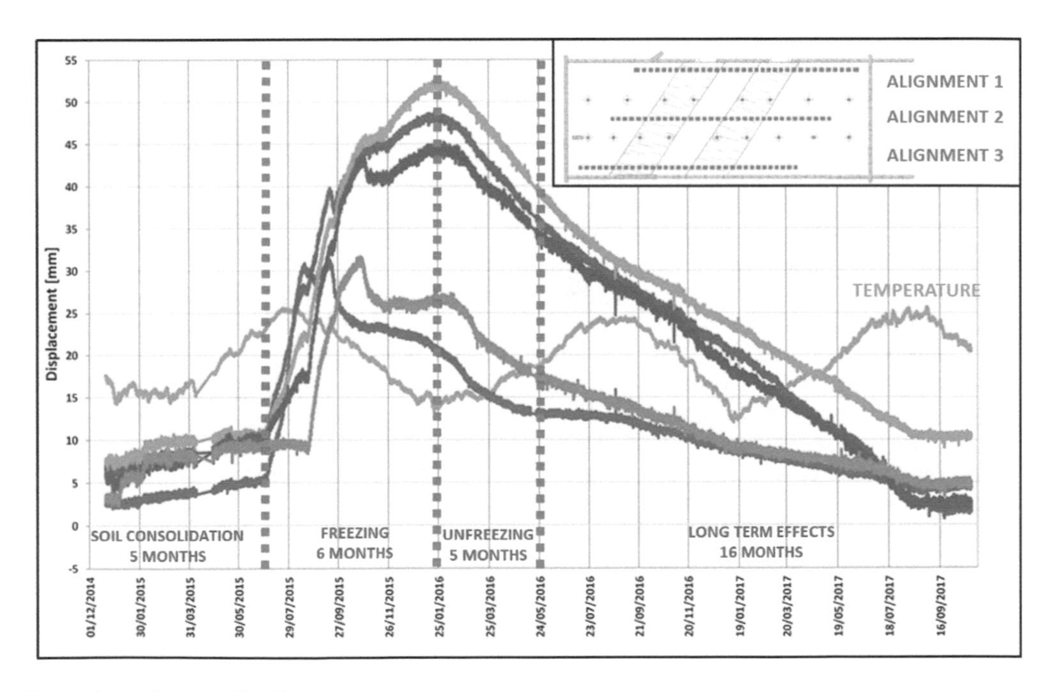

Figure 8. Alignment 2 – Settlement gauges.

9 THE ARCHAEOLOGICAL EXHIBITION

The excavation of San Giovanni Station has represented a good opportunity for important new acquisitions about the history of Rome. It has been a unique opportunity for the archaeologists because the archaeological excavations rarely reach such a depth in urban areas.

Metro C excavated about 50.000 m^3 of soil with archaeological method in San Giovanni job site and found 40.000 archaeological findings.

The excavation reached a depth of about 30 m from ground level and it crossed many archaeological strata that allowed a chronological reconstruction through the ages. The excavation phases represented a long 'time travel' that started with our times at the ground level down to the Prehistoric Times, going through Modern Age, Middle Ages, Imperial and Republican Ages. The most important structure found during the archaeological excavation was a big –35 m wide and 70 m long—reservoir of late Roman age.

Figure 9. A view of the big reservoir from the late Roma Age and a reconstruction of the area.

Due to this and other findings, the Archaeological Authority prescribed to change the San Giovanni station architectonic design in order to valorise the archaeological setting at the end of the excavation.

Accordingly, *Metro C* designed and realized a permanent exhibition of the archaeological findings in cooperation with a group of professors and architects of the Architectural Department of "La Sapienza" University of Rome.

They designed this exhibition as a long tale, telling the history of Rome using graphic elements such as images and texts printed on glass-panels, as well as some special outfitting for the exhibition of a small selection of remains selected among 40,000 findings.

Because this exhibition is inside an underground station and not in a museum, high perceptibility, fast comprehension and visual comfort were essential requirements for the design.

The most important symbol of this representation is a stratigrapher, printed on glass-panels, which consists of a vertical scale on which passengers can at the same time read their actual depth in meters from the ground level and the remarkable events which occurred at that time in the San Giovanni area and in the whole city of Rome. Also, large captions printed on the panels explain the meaning of the exhibits in some areas of the station and give the passengers the key to interpret what they see.

The horizontal levels of the station open to the public – the first and the third levels with the Line A interchange and the platform – match significant historic periods, with relevant historical events. In fact the most remarkable exhibits were found at the depth of the interchange level with the existing Line A. At this level, a big complex of hydraulic and agricultural equipment belonging to a large farm of imperial age on the edge of the town was found. The big reservoir was found in this area, so the design takes into account the outline of the perimeter walls of this structure, which define a space recognisable thanks to a different type of paving, but includes also the display of a large number of pipes and hydraulic equipment found during the excavation.

At the end of this long travel, the passengers arrive at the platform level. This level corresponds to the prehistoric times and the exhibition shows an abundant flora and a rare human presence. For this reason, at this level there are only images printed on the glass-panels.

The design and construction of this station represent a unique experience because of the exceptional synergy it expresses between architecture, archaeology and state-of-the-art in engineering.

Figure 10. Some internal views of the San Giovanni Exhibition.

10 CONCLUSION

The construction of Line C San Giovanni station in Rome has represented a work of exceptional interest for the multidisciplinary approach adopted and for the complexity and the quality of the results achieved.

The excavation has represented a good opportunity for important new acquisitions about the history of Rome. It has been a unique opportunity for the archaeologists because the archaeological excavations rarely reach such a depth in urban areas.

The procedures for building this station were heavily constrained by the archaeological remains distributed in landfill down to a depth of 18 m below ground level. These findings required an overall revision of the Line C's design in the historic centre of the city of Rome, starting from the San Giovanni station, to safeguard archaeologically sensitive strata.

The permanent exhibition inside the station is the first attempt in Italy to visualize the history of the site in the contemporary urban context with a "time travel" that started with our times at the ground level down to the Prehistoric Times

The underpassing of the existing station is an intervention of exceptional and unusual technical content, due to the surrounding conditions and constraints, and because it is inserted into a stratigraphic context that is complex and of difficult geotechnical modelling. These conditions in particular gave the freezing technique traits of absolute technical and conceptual complexity

REFERENCES

Bertero, A & Cribari, F. & Tanzi, A. & Fancello, S. & Gondolini, T. & Giannelli, F. & Lodico, M. & D'Angelo, M. & Romani E. (eds), 2016. *Metro C Rome Line C – Artificial ground freezing application for the underpass existing metro station San Giovanni Line A*; *ICGE 2016: 18th International Conference on Geotechnical Engineering, September 2016*. San Francisco, USA.

Viggiani, G.M.B. & Casini, F. & Bertero, A. & Cribari, F. & Giannelli, F. & Romani, E. (eds), 2016. *Artificial ground freezing: case studies technological development and fundamental research*; *XXIV Torino Geotechnical Conference Design, Construction & Controls of Soil improvement systems*, February 2016. Torino, Italy.

Giannelli, F. & Romani, E. & Di Mucci, G. & Sorge, R. & Bertero, A. & Cribari, F. (eds), 2016. *Rome's Line C: design and construction of tunnels under passing the Line A station with ground freezing*; *XIII International Conference entitled "Underground Construction Prague 2016*, May 2016. Prague, Czech Republic.

Romani, E. & Miniero, N. (eds), 2016. *Construction of Line C San Giovanni station in Rome*; *VIII International Symposium on "Geotechnical aspects of underground construction in soft ground"*, May 2013. Naples, Italy

Burland J.B. 1995. *Assessment of risk of damage to buildings due to tunneling and excavation*; *1st Int. Conf. on Earthquake Geotechnical Engineering*. Tokyo, 1995.

Tunnels and Underground Cities: Engineering and Innovation meet Archaeology, Architecture and Art, Volume 1: Archeology, Architecture and Art in underground construction – Peila, Viggiani & Celestino (Eds)
© 2020 Taylor & Francis Group, London, ISBN 978-0-367-46574-2

The archaeological findings are changing Amba-Aradam station design in Rome Metro Line C

E. Romani & M. D'Angelo
Metro C, Rome, Italy

R. Sorge
Astaldi, Rome, Italy

ABSTRACT: The design of Amba Aradam underground station has undergone several changes from the tender design to the construction phase. The archaeological investigations have resulted in a new positioning of the station in an area with lower archaeological risk as well as relevant changes in the construction phases and the technologies implemented. Amba Aradam station was originally placed in Largo dell'Amba Aradam in the preliminary design and it was moved twice in the following design phases. Presently it is placed in Piazzale Ippo-nio. At the beginning of the archaeological excavations, the remains of a barracks dating from the Ancient Roman period were found at a depth of 6 m below ground level. This was an exceptional archaeological finding, for both the size and the good state of conservation of the ancient structures, so that the Archaeological Ministerial Office prescribed Metro C to change the Amba Aradam design to safeguard the archeological findings at the end of the excavation.

1 INTRODUCTION

Line C is the third line of Rome underground. Once completed, it will run under the city from the South-eastern to the North-western area, over a total length of 25.6 km and 30 stations.

Metro C S.c.p.A—the General Contractor led by Astaldi—manages the construction of Line C in all its phases: design, archaeological surveys, tunnel boring, excavation and construction of stations, building of trains and start-up. At present, 22 stations and 19 km of line are open to the public.

The stretch presently under construction, named T3, extends from San Giovanni station to Fori Imperiali station, over a total length of about 2.9 km, underpassing the historical centre of the city of Rome. Amba Aradam–Ipponio station is included within this stretch and it is situated in the gardens of pIazzale Ipponio, in the Appio-Latino neighbourhood, close to the Aurelian Walls between Porta Metronia and Porta Asinaria.

The station design underwent a series of changes, starting from the preliminary design until its execution phase.

While excavating, a complex of ancient structures dating back to ancient Roman times was found, in-between the station diaphragm walls at a depth ranging from 8 to 15 m from ground level. Due both to their exceptional historical interest and their good state of conservation, the Archaeological Ministerial Office required a general revision of the architectural design of the station, so as to make sure that the archaeological remains within the future station would be available for public viewing and valorised at their best.

2 PHASE 1 ARCHAEOLOGICAL SURVEYS

The T3 stretch was repeatedly involved in archaeological campaigns during the drawing-up of the Preliminary design and of the final design.

This investigation phase, prior to the start of the construction work, is necessary to receive from the Archaeological Ministerial Office the authorisation to carry out the perimeter retaining work for the deep excavations needed to construct the underground structure: it is called "Phase 1 Archaeological survey" ("Indagine archeologica di Prima Fase").

In 1999, 2000 and 2001, Phase 1 Archaeological Surveys regarded the sites of all the structures involved in the T3 stretch. Later on, starting from 2006, the General Contractor was given by contract the responsibility to complete the archaeological surveys, undertaking 7 sampling surveys. Two of them, both related to Amba Aradam station, are located respectively on the corner of Via dell'Amba Aradam and Via di Villa Fonseca and in Largo dell'Amba Aradam.

Two more archaeological surveys performed in Piazzale Ipponio and within the gardens of Via Sannio provided some important information which has affected the design of the T3 stretch.

The sampling surveys carried out in Piazzale Ipponio showed heterogeneous underground installations in the first 8.50 m layer, resulting from the made ground deposited in the 1930s. Under this level emerged a layer characterized by late Middle Ages ceramic materials. At the depth of 10.50 m from ground level the excavation was interrupted for security reasons, due to the interception of the aquifer. The surveys continued by drilling two archaeological boreholes, which revealed a series of anthropic backfills down to about 15–18 m from ground level.

The sampling surveys carried out along Viale Ipponio as well as 45 archaeological boreholes drilled before the start of the construction in the volume of the future station determined the depth of the upper interface of the Late-Antique stratigraphy at about 11 m from ground level. The results of the archaeological surveys of the first phase carried out since 1999 have enabled the General Contractor to draw up the document called "Archaeological Surveys Handbook" ("Prontuario delle Indagini Archeologiche"). This document established with the Archaeological Ministerial Office the excavation methods which must be applied during phase 2 Archaeological Surveys, starting from the design stage, so as to quantify the resources to employ and subsequently estimate the time necessary to complete the archaeological excavations. Phase 2 archaeological surveys are being carried out at the same time as the underground works.

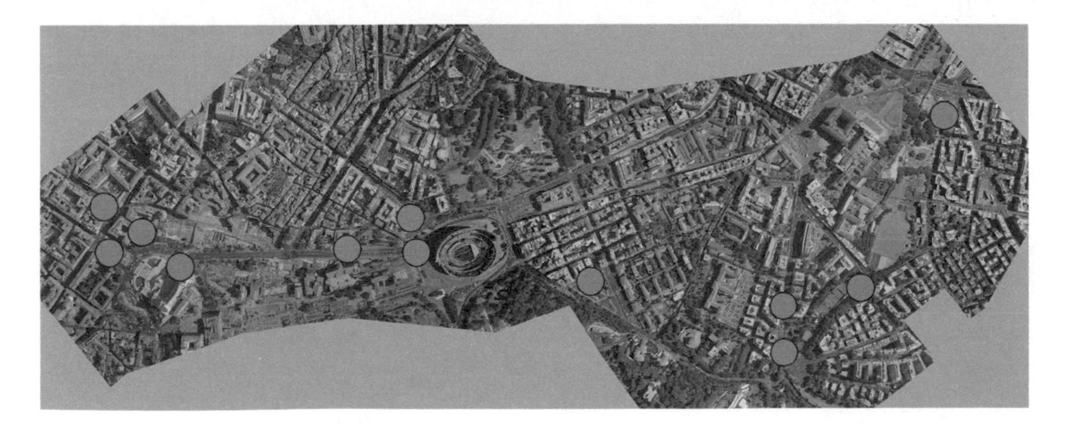

Figure 1. The archaeological investigations along T3 stretch.

3 GROUND CONDITION

The geognostic surveys carried out identified the following stratigraphic succession of the area in question. Its identification is extremely useful to interpret the complex hydrogeologic setting in which Amba-Aradam-Ipponio station is situated. Starting from ground level, the following lithotypes can be found:

- Complex of the made ground, 11 to 19 m thick: it is mainly made of backfills and of the deposits resulting from historical floodings, consisting of sandy-silty sediments as well as fragments of bricks, ceramics and masonry.
- In the area of Piazzale Ipponio, the thicknesses of the recent floods of the Tiber River vary considerably starting from Via Illiria, where they amount to about 2.5 m, to Via dei Laterani, where they exceed 13 m. Such floods are characterised by both horizontal and vertical variability in terms of clayey, silty and sandy content.
- The AR unit of the paleo-Tiber rests beneath the alluvial cover. These sediments present variable thicknesses along the body of the station, since they were locally eroded in a massive way by the ancient Fosso dell'Acqua Mariana. The unit consists of clayey silts and silty clays feably sandy.
- The SG deposit of the paleo-Tiber lies at a depth of about 28 m from ground level. This unit consists of medium-fine heterometric gravels with an abundant sandy matrix and a thickness of 7 to 12 m, down to a maximum depth of -39/-40 m.
- The bedrock is made up of the Pliocene deposits: the main characteristic of the unit is the high degree of consolidation it underwent in the course of time. They are silty-clayey deposits with frequent sandy horizons. The deposit is situated at elevated depths; the formational roof is found at 35 ÷ 40 m from ground level.

The piezometric measurements in the various strata of the soil have identified two distinct aquifers: the first one is phreatic (within the strata of the alluvial and backfill soils, whose level is situated at about 8–10 m from ground level), while the second one is artesian, within the stratum of gravels and sands of the Pleistocene deposits, situated at about 16 m from ground level.

4 THE CHANGES FROM THE PRELIMINARY DESIGN TO THE FINAL DESIGN

The Preliminary Design of Line C under the tender established the construction of the stretch between Lodi station and Amba Aradam station in a soil stratigraphy presenting a backfill cover of about 20 m from ground level. In the following picture is shown the original altimetric profile of the line where the state of the backfills is clearly marked in yellow.

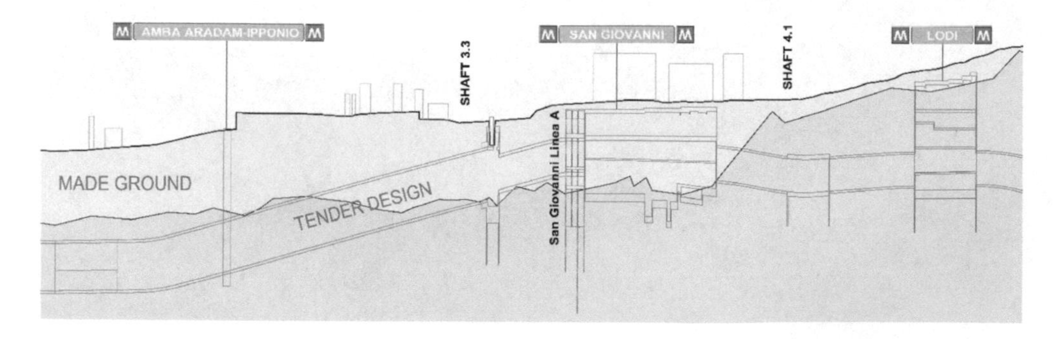

Figure 2. Longitudinal profile of Tender design between Lodi and Amba Aradam Station.

The Preliminary Design of Amba Aradam station established the construction of a rectangular structure of modest planimetric dimensions – 45 × 15 m – down to the depth of about 30 m, under the gardens of Largo dell'Amba Aradam, close to Porta Metronia. The main shaft, situated in the middle of the line galleries, comprised the technical installations area as well as part of the staircase system which links the platform level to the atrium, located at about 10 m from ground level. Starting from that level, a series of underground narrow tunnels which interfered with the archaeological strata branched off towards the 3 entrances. Moreover, from the platform level started one more underground tunnel which included the second staircase.

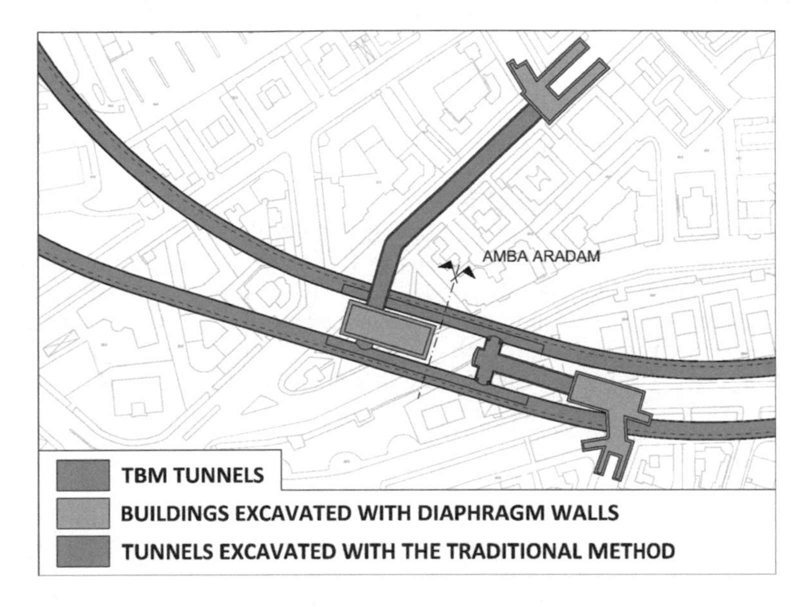

Figure 3. Tender design: Layout of Amba Aradam Station.

Given the aquifer was at about 8–10 m from ground level, the construction techniques applied in this design included the widespread use of the soil consolidation by the method of jet grouting, both to waterproof the excavation in the shafts and to consolidate and waterproof the soil during the construction of the underground tunnels linking the platforms to the surface.

As previously described, preliminary excavations and surveys have been carried out along the T3 stretch through archaeological boreholes in various areas since 1999, in order to preventively verify the possible presence of ancient structures. While excavating, wall structures emerged in the backfills, as can be seen in the following picture, representing the area of Porta Asinaria.

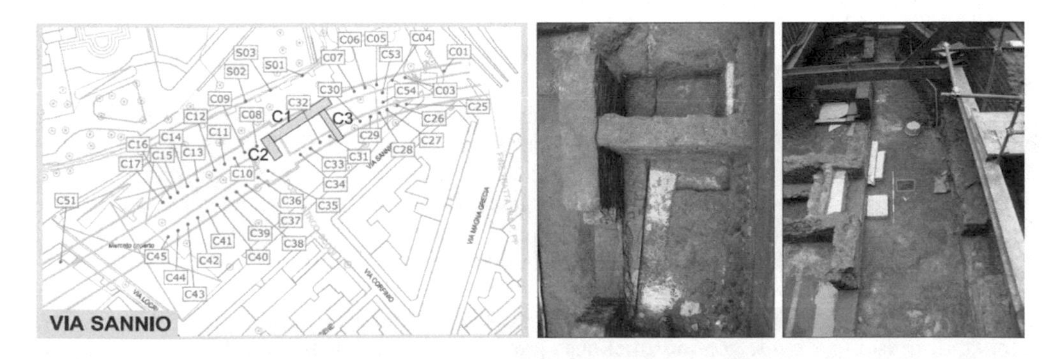

Figure 4. Archaeological investigations at Porta Asinaria.

Following these surveys, the Archaeological Ministerial Office prescribed that the tunnel should be excavated by the open excavation method for the stretch which extends from the middle of Via La Spezia up to Largo Amba Aradam by archaeological method, down to the sterile ground, located at a depth of about 18–20 m from ground level, because such excavations involved some backfill strata at very high archaeological risk. Besides, the possibility of using jet-grouting was excluded, as this technology was considered "...extremely invasive, destructive, and therefore incompatible with the areas characterised by complex remains of archaeological interest".

These prescriptions resulted in the need to study a new design solution to solve the problems related to the entire track in question. The design variation regarded the track between the Amba Aradam and the Lodi stations for about 2 km, including also San Giovanni station, an important interchange with the existing Line A and two ventilation shafts.

That resulted in an approach to the design and the definition of the interventions totally different from what had previously been planned, with important consequences on all the design choices.

In order to solve the archaeological issue mentioned above, the solution adopted was to abandon the restriction represented by the existing set-up within Line A San Giovanni station, through the necessary altimetric variant, so as to underpass the existing building. In such a way, the line galleries could be placed entirely in the archaeologically sterile soil, avoiding the massive made ground covers and limiting the interference with these strata to the station structures only. This solution radically solved the issues related to the archaeological interferences of the line galleries, which were only limited to the station and the ventilation shaft.

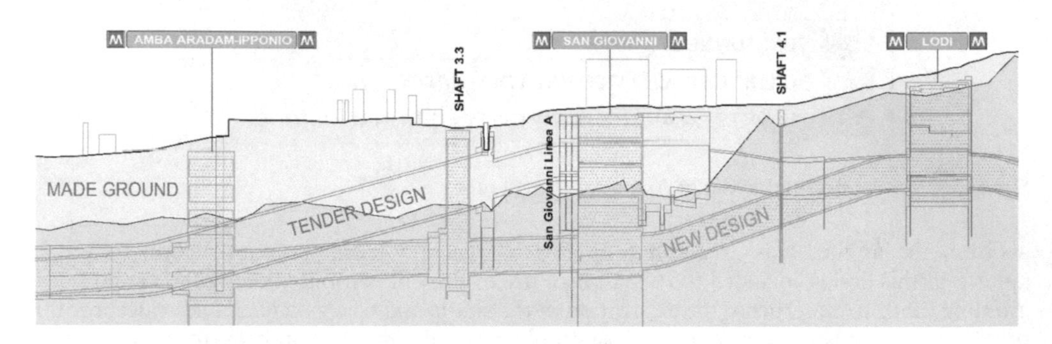

Figure 5. Longitudinal profile with a comparison between Tender design and the new design.

As far as Amba Aradam station is concerned, the Archaeological Ministerial Office prevented the use of jet-grouting. However, the station originally designed in Largo Amba Aradam largely implied the use of that technology, given the several narrow underground tunnels to be excavated under the water table. As a consequence, the station was relocated to a position suitable for its box-shaped structure, entirely confined within its perimeter by diaphragm walls fitted at the base in the impermeable strata, so as to construct a water-tight building with no need for soil consolidation.

To this end, a topographic comparison study was performed by professor Funicello, overlapping the cartographies available for the last 100 years, that is the I.G.M. of 1907–24 and the papers of the Comune di Roma (Cartesia 2000). This study provides a representation of the positive and negative variations of the soil level, showing the anthropic activities resulting from excavations and backfill. (brown colored in Figure 6).

This representation clearly shows the rise of the made ground in the external area of the Aurelian Walls from Via Sannio to Piazzale Ipponio and beyond. In particular, it can be clearly seen that the area of Piazzale Ipponio, where Amba Aradam-Ipponio station was relocated, presents a made ground thickness 8 to 12 m thick, which should rule out the possibility of finding archaeological remains in those strata, or at least keep it very low.

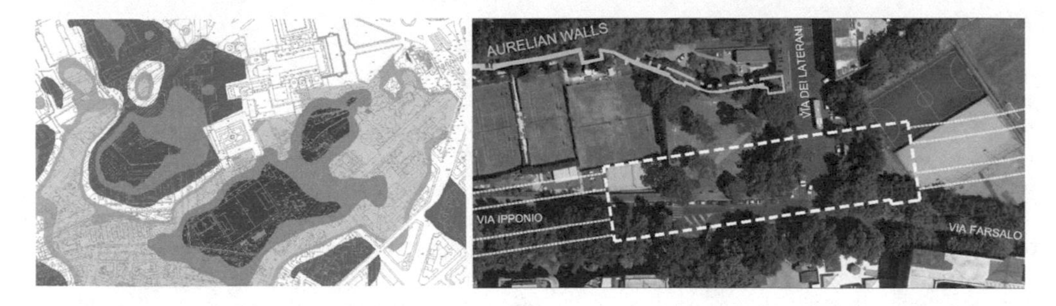

Figure 6. New localization of Amba Aradam Station.

A further change to the design resulted from the need to carry out considerable archaeological excavations down to a depth of 18–20 m from ground level without using the usual top-down method within these strata.

The archaeological excavation requires a uniform progression on the entire area of intervention from top to bottom (top-down), studying and recording the stratigraphic succession by horizontal levels as belonging to the same age. The top-down excavation method implies the realization of intermediate slabs during excavation which act as a support for the perimeter diaphgram walls using the soil as supporting element during their construction.

This method is not suitable for the archaeological excavation, because once constructed an intermediate slab, the excavation moves on through limited operative openings designed for the purpose. Through them, the soil is gradually removed by mechanical means, progressing horizontally without any possibility to safeguard the archaeological remains and study the stratigraphic succession at the same time in the entire area of interest.

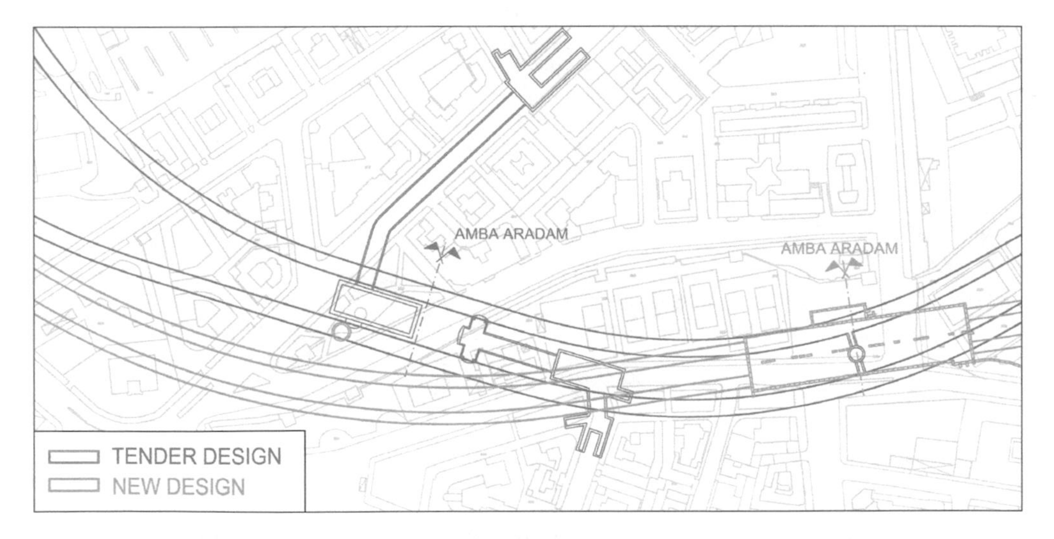

Figure 7. Plant with a comparison between Tender design and the new design.

5 THE NEW DESIGN OF AMBA ARADAM – IPPONIO STATION

The new Amba Aradam - Ipponio station is situated within the gardens of Piazzale Ipponio on the corner between Via Farsalo and Via dei Laterani. It is delimited to the North by the Aurelian Walls, which extend along Via della Ferratella, and it is set in a urban area characterised by residential buildings.

The station has a rectangular plan of external dimensions of about 120x30 m and reaches a depth of about 30 m from ground level. The construction of an underground structure of such dimensions enables the entire body of the station to be contained in it, so as to avoid excavating the long underground tunnels that would interact with the archaeological strata.

The main body consists of 4 underground levels/floors apart from the slab roofing and the foundation slab. Three out of four are open to the public, while the fourth one is a level totally dedicated to the technical installations of the station. The station can be accessed via three separate entrances, situated along the three main directions: Via Farsalo, Via dei Laterani and Viale Ipponio. Once in the hall and overcome the line of the turnstiles, one can get to the platform level through three bodies of escalators and stairs passing through the 2 technological levels.

The excavation is retained between diaphragm walls 1.20 m thick and about 45 m high from ground level so as to reach the water-impermeable stratum of the Pliocene clays.

The executive phases and the technological choices for the construction of the station took advantage of the recent experience of the excavation of San Giovanni station, in which a new methodology called "Modified Top – Down" was developed.

Such a methodology allows to benefit from the "top – down" method, safeguarding the pre-existing archaeological structures. Once constructed the perimeter diaphragm walls, the excavations can start by archaeological method down to a depth of about 3 m below the intrados of the slab covers. Subsequently, once constructed this slab, the intermediate ones were excavated and constructed down to the foundations, deepening the excavation at each step of at least 3 m with respect to each slab intrados. This deepening guarantees the step by step progression of the excavation from top to bottom for the further deepening with a horizontal development, which is typical of the excavation by archaeological method.

Therefore the slab covers and the atrium which are situated within the archaeological layer cannot be realized using the soil as temporary support, except for limited portions. As regards

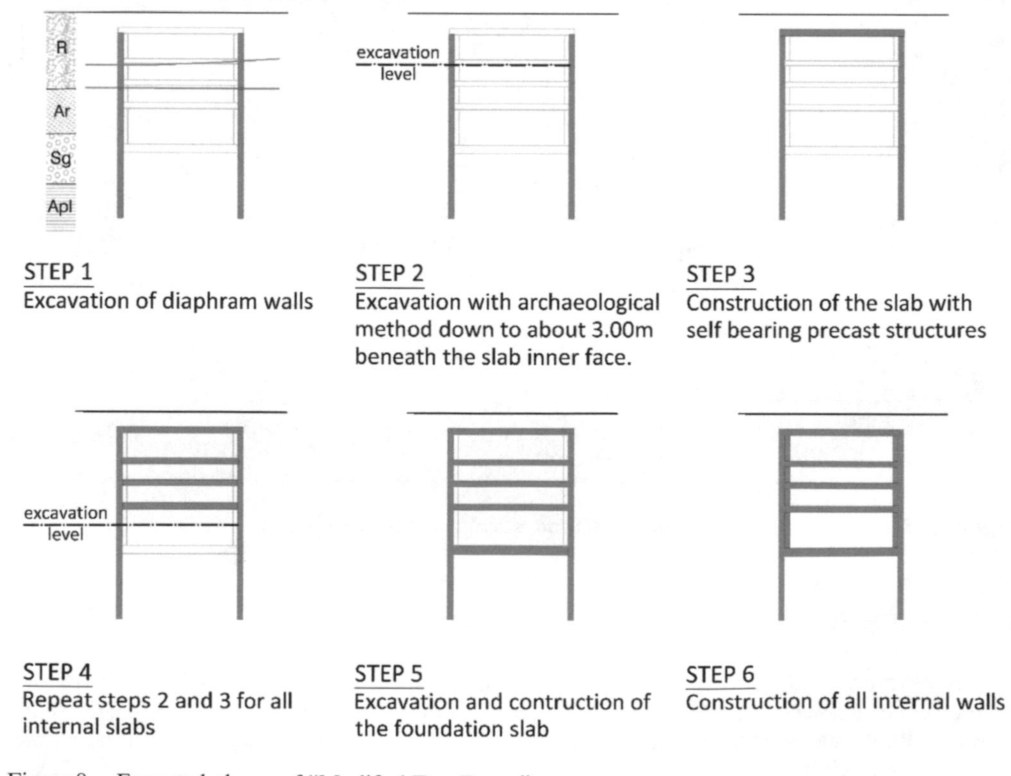

Figure 8. Executed phases of "Modified Top-Down".

these structural elements, self-supporting prefabricated segments were chosen, to limit the use of temporary elements such as timbering as much as possible.

The slabs of the first technical level and of the mezzanine floor, instead, being situated beneath the archaeological layer, were made of steel reinforced concrete structures, using the soil as a temporary supporting element, before the maturity of the cast-in-place concrete.

6 THE ARCHAEOLOGICAL FINDINGS AND THE FURTHER PRESCRIPTIONS

During the second phase of the archaeological excavations, carried out in-between the perimeter diaphragm walls of the station at a depth of about 9 m from ground level, a unitarian complex of structures was found. It supposedly dates from the first half of the II century AD, specifically to the Age of Hadrian, and it can be functionally identified as a barracks -(castra).

The complex takes up the entire Southern longitudinal centreline of the station body and it extends beyond its edges. The structural complex presents all the elements which define the peculiar function of the barracks: a series of 39 rooms built in a unitarian construction phase on an area of about 750 m^2 Some of them present mosaic pavings and frescoed walls, on the two sides of a long corridor which goes beyond the perimeter diaphragm walls of the station.

The excavation deepening on the Northern side of the station has brought to the finding of several more rooms. Some of them were richly decorated and were found at a depth of about 3 m under the barracks. The bottom level structures consist of two distinct units: a lodging area and a functional section, separated by a a large open area.

The lodging area, called the "Commander's Domus", is characterised by the remains of an important building which presents wall frescoes as well as stone, mosaic and fictile pavings in very good conditions.

The architectonic structure dating from the Age of Hadrian was modified and enlarged in the course of time until the complex was abandoned, in the second half of the III century AD, at the same time as the construction of the Aurelian Walls. The evidence of this is the fact that the lodging area is level with the grade plane of the Walls.

The archaeological findings are delimited to the East by a massive wall, about 1.60 m thick and 3.30 to 4.50 m high, which extends for about 28 m and takes up the entire width of the station body. A limited portion of the structure presented a painted grout lining.

To the East, beyond the massive wall which surrounds the built-up area, there used to be a terraced garden area, sloping towards the Acqua Crabra Ditch.

These massive ancient structures were found at a depth interfering with the first two underground slabs of the station and this entailed further changes, as the works progressed, to the structural design and to the succession of the executive phases. Among them, the most important was certainly the replacement of the atrium level with a temporary steel structure which allowed to effectively support the perimeter diaphragms while excavating, maintaining the ancient structures of the barracks in situ, up to the reaching of the wall foundations.

Because of this exceptional historical-archaeological finding, as well as the good state of conservation of the ancient structures, the Archaeological Ministerial Office required an overall revision of the architectural design of Amba Aradam—Ipponio station. The authorisation to the temporary disassembling of the structures found was bound to their subsequent relocation, to ensure their maximum safeguard and their enjoyment and valorization within the future station, creating a museum exhibition, functionally separated from the station in operation.

Figure 9. Archaeological findings in Amba Aradam Station.

Figure 10. Temporary steel structures installed in Amba Aradam Station.

7 THE CREATION OF THE MUSEUM AREA

Starting from the prescriptions of the Archaeological Ministerial Office, the new design solution aims at developing a project which can integrate and valorise the ancient structures within the contemporary context.

Such a solution involves a spatial, functional and operational separation of the premises dedicated to the archaeological exhibit from the area dedicated to the station. However, the two functions keep a visual continuity thanks to the glass-windows inside, which keep the station apart from the "new" museum space.

The design involves the creation of a large open-air lowered area which enables light and air to get to the underlying levels of the underground and the museum.

The museum can be accessed through the square. Once in, the visitors can make a complete visit of the archaeological site from above by walking on a suspended catwalk, which allows for a closer look at the relocated barracks.

Figure 11. Internal views of the new Museum close to Amba Aradam Station.

REFERENCE

Egidi, R. & Filippi, F. & Martone, S. 2010. *Archeologia e Infrastrutture. Bollettino d'Arte – Volume Speciale, Ministero per I Beni e le Attività Culturali.* Springer, Berlin.

Tunnels and Underground Cities: Engineering and Innovation meet Archaeology, Architecture and Art, Volume 1: Archeology, Architecture and Art in underground construction – Peila, Viggiani & Celestino (Eds)
 ISBN 978-0-367-46574-2

Underground car park in the ancient "Morelli" cavern in Naples

F. Rossano, A. Bellone & M.A. Piangatelli
CIPA S.P.A., Rome, Italy

ABSTRACT: In 2004 works started for the now completed seven-level automated car park accommodating 480 parking spaces inside the pre-existent "Morelli cave", located in a strategic area in Naples. Since 470 BC Greeks initiated the growth of the fascinating world of the underground Naples, of which this cave belongs. It seems that once here the god Mithras was worshipped. it's also become a quarry for tuff extraction and in the 1600s it became an aqueduct serving the area. In 1853 the Bourbon Tunnel, a project commissioned by King Ferdinand II of Bourbon to allow easy escape from the Royal Palace in case of riots, was connected to this network of tunnels. The cavern became an air raid shelter during the Second World War, then a deposit of cars and motorcycles in the '50s and' 60s, a disposal area and it has been abandoned in the last 40 years before it's new utilization.

1 INTRODUCTION

Over time, the overcrowding and congestion of major cities' historic centres has led to seeking spaces underground that can accommodate activities and car parks, as in the case of the Morelli cavity, site of the future car park by that name.

Before describing its characteristics, design solutions, and executive problems, it is necessary to introduce the site historically into the broader landscape of the cavities present below ground in the municipality of Naples.

The province of Naples, which includes the area of the Municipality of Naples, has a large number of cavities. The existence of "Naples underground" is connected with the morphological and geological conformation of its territory, consisting of tuff stone which has entirely special features of lightness, brittleness, and stability.

The first transformations in the territory's morphology, taking place thanks to the Greeks starting from 470 BC, initiated the growth of the fascinating world of Naples underground. These transformations were dictated by water provision requirements, which led to the creation of underground cisterns to collect rainwater, and by the need to recover construction material to erect the buildings of Naples. In subsequent centuries, the city's expansion led to the construction of a proper aqueduct, which allowed drinking water to be collected and distributed thanks to a series of cisterns linked to a dense network of tunnels. During Roman rule, the existing aqueduct was expanded and perfected, but with the rise of the Angevins in 1266, the city saw great urban expansion, which obviously corresponded with increased extraction of tuff from the subsoil to construct new buildings. This confirmed a particular feature of Naples: that of having been generated from its own innards, with buildings rising directly above the quarry that provided their building material.

Decisively impacting the destiny of Naples underground were some edicts between 1588 and 1615, which prohibited introducing building materials into the city in order to stem Naples's uncontrolled expansion. To avoid punishment while still meeting the need for urban expansion, the citizens had the idea of extracting the tuff beneath the city, by using pre-existing shafts, enlarging the cisterns designed to contain drinking water, and digging out more. Extraction of this kind, which took place from top to bottom, required special techniques in order to guarantee the stability of the subsoil and prevent undesired collapses.

Only in 1885, after a terrible cholera epidemic, was the use of the old water distribution system abandoned in favour of adopting the new aqueduct, which is still in operation.

The most recent underground project dates back to the Second World War, when to offer safe refuge to the population, the decision was made to adapt the ancient aqueduct's structures to the citizens' needs. Throughout Naples, 369 shelters in the caves and 247 collapse-proof shelters were set up. An official list from the Ministry of the Interior from 1939 lists 616 addresses that led to these 436 shelters, some of which with more than one access. The arrangement of the shelters led to an additional splitting-up of the ancient aqueduct.

At War's end, given the dearth of means of transport, almost all the rubble was unloaded into the subsoil, almost as if wishing to bury with it all the memories of that sad period. Until the late 1960s, there was no more talk of Naples underground, even though many went on using the shafts as dumps.

However, starting in 1968, some instabilities began to occur, due essentially to ruptured sewers, or leaks in the new aqueduct. In all the world's cities, these problems are seen when sewage gurgles to the surface, or with flooding; but in Naples, precisely because of the vast, hollow subsoil, the same problems become evident with large sink holes. This is due to the fact that aqueduct water or sewage almost always finds a preferential path towards the old shafts, where all those incoherent materials at the roof of the tuff and surrounding the leaks liquefy. This causes a void, which proceeds upwards and becomes evident only when the final layer – consisting of slabs of terrain or of roadbeds– collapses into it.

Coming to the subject hereof, the "Morelli cave" (see Figure 1) is located in one of the most strategic areas of the city of Naples (see Figure 2), just a few hundred metres from the downtown Piazza Dei Martiri, Villa Comunale, and Via Caracciolo; a "Bourbon-era" tunnel, which in the Neapolitan circulation plans will connect with Piazza del Plebiscito and the Via Toledo area, departs from it.

For years the car park was used on "on grade level" at about 9.00 m above sea level, with extremely limited capacity.

However, the project developed a total of 435 spaces, 225 of which for rotation parking and 210 for private garaging. It bears mentioning in any event that the original plan called for 673 car spaces, and many of them (no fewer than 238) were sacrificed in drawing up a design that respected the safety and stability needs, as well as the state of the locations, to the maximum extent possible.

The general layout of the car park located within the Morelli cavity (see Figure 3) consists of a structure in reinforced concrete cast onsite, and is organized in 14 modules, seismically jointed together.

Figure 1. Pictures of the "Morelli" cavern.

Figure 2. "Morelli" cavern location into the urban context of Naples.

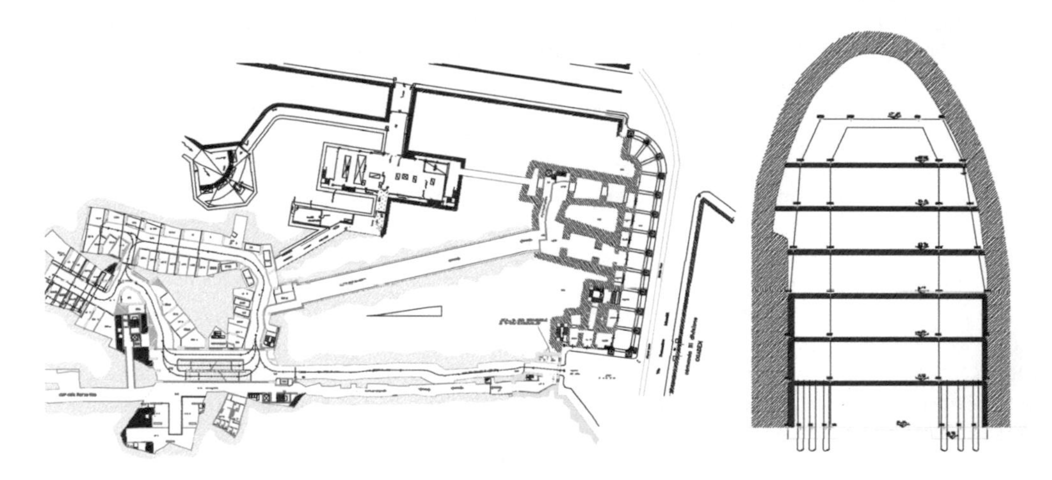

Figure 3. General layout of the parking and a typical section.

Each module has a total of 6 decks in elevation and a foundation with plinths and a connection slab – except for the smaller ones to be done in rather special points in the cavity, such as niches and enclosed spaces between the existing masonry works. A newly-built tunnel serves the car park, with an entrance from the arcades of the building at Via Morelli. This tunnel, which entailed considerable difficulties connected essentially with having encountered additional cavities – often full of detritus – along the route, has a mezzanine that allows pedestrians to access the structure.

2 STRATIGRAPHY OF THE SUBSOIL

The Morelli cavity develops entirely in the setting of the tuff mass of the Monte Echia promontory, where the presence of a series of buildings over it in fact hampered the execution of systematic probes from above to define the tuff roof's development in detail.

The stratigraphy of the subsoil was thus gleaned both from direct observations conducted on the walls of the Morelli cavity, and from the results of the ongoing core sampling, which was possible to perform both in the same cavity and in some yards above it. Based on these observations, it may be said, with a sufficient degree of approximation, that the tuff coverage is no less than about 4.0 m with respect to the highest part of the cavity.

Above the tuff mass, practically to ground level, are loose pyroclastic soils and, to a lesser extent, landfill.

The free surface of the groundwater, based on some observations made in the car park area, is at about +1.5 m above sea level.

3 GEOTECHNICAL CHARACTERIZATION OF THE SUBSOIL

3.1 *Degree of fracturing of the mass*

The degree of fracturing of the rock, as obtained from the examination of the various probes performed in the Morelli cavity's tuff walls, is generally rather contained. Rather high RQD (Rock Quality Designation) levels were in fact obtained, mostly between 90 and 100%; only in a few cases, near individual discontinuities, was an RQD value of less than 50÷60% found.

Moreover, as far as is visible to the naked eye inside the Morelli cavity, there are no real families of discontinuity, but only localized cracks. However, it cannot be ruled out that in some areas, the mass, on the inside, may in fact be characterized by a higher level of fracturing than appears from an examination of the exposed surfaces, or that there might not be discontinuities parallel or sub-parallel to the cavity walls, produced by the release of tension due to the making of the cavity itself.

3.2 *Physical-mechanical properties of the soils and rocks*

For the purposes of the performance and the stability of the works in question, the loose pyroclastic soils present above the roof of the tuff formation are doubtlessly of little importance, in both geotechnical and applicative terms. As for the tuff, on the other hand, it bears mentioning that in the case in point, it is of the utmost importance to define its resistance characteristics not only at the level of the individual samples, but also and above all on the scale of the mass.

Starting from the results of the laboratory tests as reported in the 2001 document drawn up by the Department of Geotechnical Engineering (Dipartimento di Ingegneria Geotecnica – DIG) of Università di Napoli Federico II, it is found that the weight of the unit of dry volume equals on average 10.8 kN/m^3. Depending on the natural water content, equalling 0.20, a weight of the unit of volume of the tuff equal to about 13.0 kN/m^3 is obtained.

The average value of the breaking strength is approximately 3.5 MPa for samples subjected to testing after having been air-dried, and 2.6 MPa for samples subjected to testing in saturation conditions. It is then observed that the saturation leads to a reduction in resistance equal to about 25%, while in both cases the standard deviation is about 0.7 MPa. Based on statistical considerations, one may then take into account a characteristic value (obtained by subtracting twice the standard deviation from the average value) of the uniaxial compressive strength equal to about 2.1 MPa for the dried samples, and 1.2 MPa for the saturated samples.

As for the breakage envelope and the relative parameters c' and ϕ', the results of the triaxial CID compression tests performed at DIG, Università di Napoli Federico II, were taken into consideration. In the σ'_1-σ'_3 plane, the values of the breaking strains obtained by the tests performed on saturated samples, with the exclusion of some clearly anomalous results, yielded an effective cohesion value of 0.74 MPa and a 17° angle of friction.

However, it is stressed that this angle of friction value is far below the typical range of values obtained for yellow Neapolitan tuff in the abundant experimentation performed at DIG, Naples. Based on these tests, in fact, in may be said that the angle of friction always exceeds 20°, with typical values between 24° and 30°.

The first one obtained is thus to be considered an anomalous value which is probably influenced by the limited number of tests performed and the strong heterogeneity of samples tested, as proved by the marked variability of the uniaxial compressive strength found among the tested samples.

The tuff's resistance parameters are cross-correlated with the negative degree of correlation (Aversa 1989); that is, increased cohesion often corresponds with a diminished term of friction. Since the stability of the excavations depends highly on the cohesion value, the preference was to assume a higher value (25°) for the angle of friction and to obtain the cohesion with the compressive strength characteristic of the dried material, thus obtaining c'_k = 0.66 MPa. On the other hand, the analogous value for saturated tuff equals about 0.40 MPa.

As concerns deformability, the uniaxial tests also yielded a Young's modulus between 1500 and 3500 MPa.

As rock mechanics tells us, to carry out the mechanical characterization of a mass, once the properties of the material of which it is made are known at the scale of the samples used in the lab, it is necessary to assess certain factors that influence and determine their behaviour on a larger scale. Of these, the most important ones are the so-called "size effect" and the degree of fracturing of the mass.

In particular, *"size effect"* is essentially linked to the material's inhomogeneity and to the possible presence of micro-cracks. In this regard, special laboratory tests have clearly shown that with increased sizes of the sample subject to rupture testing, the cohesion and the elastic modulus tend to diminish to an asymptotic value. To take this diminution into account, in the case of yellow tuff a reductive coefficient equalling about 0.7 ought to be considered.

The *degree of fracturing*, however, is linked mainly to the development of the cooling and cementation processes of the loose pyroclastic material from which the tuff originated, as well as to later seismic and/or tectonic phenomena and, in the final analysis, to states of overstressing caused by humans, which may have caused the rock to break locally. The influence of the degree of fracturing on the overall behaviour of a rock mass is extremely important: one need only consider that in the presence of two or more families of discontinuity, each with an interaxis between cracks on the order of one decimetre (a rather frequent case, for example, in highly tectonized rock), the behaviour at excavation is no longer that of a rock, rather than that of a gravel, or at any rate of a fundamentally non-cohesive soil.

Aside from this extreme case, rock fracturing results in a reduction both of cohesion and of the elastic modulus, which of course grows more accentuated as the rock's integrity lessens. In the specific case, the high RQD values found with the investigations performed result in reductive factors that may be estimated at about 0.85÷0.90.

Therefore, in conclusion, the tuff mass in question may be characterized with a constitutive elastic/perfectly elastic rule, with the criterion of Mohr-Coulomb yield point and non-associated flow rule, or an elastic-plastic model with hyperbolic type tension/deformations bond, using the average values of the parameters in Table 1.

Table 1. Average values of the tuff parameters.

Parameters	tuff above the water table	tuff below the water table
Weight of unit of volume	14 kN/ m^3	14 kN/ m^3
Angle of friction	25°	25°
Effective cohesion	350÷400 kPa	250÷300 kPa
Tensile strength	70÷100 kPa	40÷60 kPa
Elastic modulus	1000 MPa	800÷1000 MPa
Poisson's ratio	0.25	0.25
Dilatancy angle	0	0

4 DESCRIPTION OF BEARING STRUCTURES

The vertical structures consist of septa 40 cm thick and variable in length, placed, for the first useful elevations, crosswise against and longitudinally with the cavity's axis of development. They are conceived as collaborating with the tuff walls up to an elevation of +9.65 m; for the remaining parts as well, the septa are connected to the very walls of the cavity.

4.1 *Deck slabs*

The slabs are to be understood as cast full slab on site, with two-way reinforcement, for a total thickness of 35 cm.

4.2 *Foundations*

The foundations consist of plinths 80 cm thick, in some cases on piles 400 mm in diameter and 6.00 m in length above the plinths' level of installation. The plinths are joined to a connection slab about 40 cm thick, and upper surface coinciding with the upper surface of the plinths. The indicated upper surface elevation is set at +0.50 m.

As regards only the boreholes, the piles were made from the work site elevation, appropriately levelled to +3.20 m, except for the already-indicated piles/piers, which instead shall be complete with reinforcement and casting up to the borehole mouth level, as they had have to function as provisional support for the first three slabs.

5 CONSTRUCTION PHASES

The works to build the Morelli car park began in June 2004. Hereunder, the modes of operation and design for each working process shall be described, highlighting the problems faced in so particular a site, which have prolonged the working operations.

Working operations began with the bench and stabilization of the walls and of the cavity crown.

This stabilization was done through the insertion of structural elements, threaded bars and Ct-Bolts, suitable for improving the geomechanical characteristics of the mass, and whose characteristics are to be analysed in greater depth below.

Figure 4. Lattice grids that fit the crown.

An additional intervention was carried out in correspondence with the cavities' crowns. In some sections, these were reinforced with lattice ribs (see Figure 4), constituting a kind of lining; of course, since the cavities, once the excavation is done, reach heights up to a maximum of 30 metres, for the installation of these ribs, as well as for the stabilization on its own, scaffolding was erected ad hoc, with irregular geometries adapted to the cavity's morphology, making it possible to reach these heights (see Figure 5). The operations were completed with the installation of wire mesh and shotcrete.

Figure 5. Part of the scaffolding.

6 CAVITY STABILIZATION: CT-BOLT

The cavity was consolidated, in accordance with precise needs, using various techniques, such as appropriately grouted Dywidag anchor bars, but particularly worthy of mention is the additional system known as CT-Bolt, one of the systems most widely used to support and consolidate rocky walls (see Figure 6).

The purpose of a system conceived in this way is to contain the deformation phenomena and, in the event there are disarticulated prisms of rock, to support them.

Of course, the choice of anchoring type is closely linked to the type of rocky mass and, in addition to the specific work site conditions, to technical and economic considerations. CT-Bolt is a corrosion-resistant anchoring that is installed following the geomechanical principle of temporary and immediate rock support using an expansion shell and then grouted to give it permanent anchoring characteristics. This system presents considerable differences in technique, performance, and cost (of the element and of its installation) from other types of anchoring. A rigid sleeve in polyethylene placed between the steel bar and the rock guarantees the seal, eliminating the corrosive action of water, and is at the same time used as grouting conduit.

The sleeve's particular shape also acts as a continuous centring device for the bolt in the borehole, keeping water from coming into contact with the steel and guaranteeing an anti-

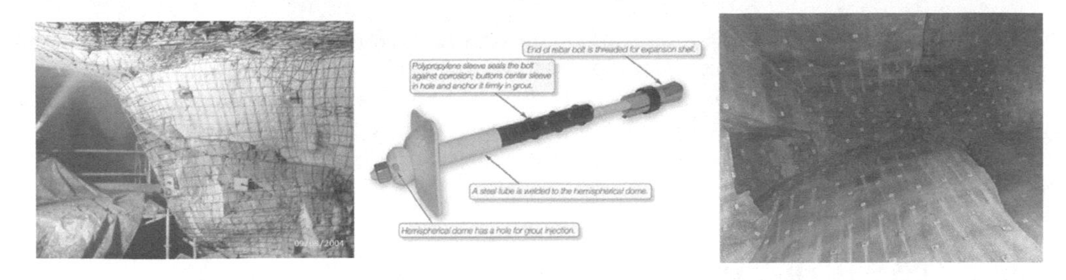

Figure 6. CT-Bolts on the crown before they are covered by a shotcrete layer.

friction effect against slippage. The CT-Bolt comes with a dome-shaped body that has two functions: it supports and distributes the load on the bolt plate and serves as a grouting chamber.

The grout is pumped into the mixing dome and through the polyethylene sleeve; at the upper end it flows between the sleeve and the rock until escaping from the bolt plate. The characteristics are shown in Table 2.

Table 2. CT-Bolt characteristics.

Model	Bar diameter	Ultimate tensile strength, point anchored	Ultimate tensile strength, fully grouted
C-TUBE M20	18.6 mm	150 kN	180 kN
C-TUBE M22	21.6 mm	250 kN	318 kN
Standard length		From 1.5 m to 8.0 m (others upon request)	
Boring diameter		From 45 to 51 mm	

7 BUILDING OF THE MORELLI TUNNEL

The excavation of the entrance tunnel to the Via Domenico Morelli car park was done first in full face, supporting the excavation cross section with bolts and the installation of electro-welded mesh (see Figure 7.).

Then, in September 2005, intercepting a cavity filled with loose debris required changing the mode of excavation; in fact, after having grouted with lean concrete to allow the cavity to be filled after stabilizing the excavation face with shotcrete, the decision was made to excavate a truncated

Figure 7. Excavation done by rotary cutter and supported by electrowelded mesh and CT-Bolts.

conical cross section with double forepoling on the crown, reinforcement pipes diam. 139.7, 8 mm thick, with an interaxis of approximately 25 cm, to allow this problem to be overcome.

Subsequently, the first-phase lining was made (ribs and shotcrete), as well as sealing, formwork reinforcement and the casting of the final lining, later completed in June 2006 with the grouting of an intermediate slab allowing the upper portion of this tunnel to be used by pedestrians.

8 BUILDING OF THE STRUCTURE IN ELEVATION

The building of the structures constituting the car park began in July 2006 from module A, continuing with the construction of modules I and H, B and G; for the other modules, the plan was to proceed alternately to the current cavity entrance from Largo Morelli.

This is all to build a encirclement of the central pier in natural tuff which characterizes the cavity itself.

To pursue this purpose, for modules A, B, B1, I, H, G, and F, the floors at elevations +6.60 m, +9.65 m, and +3.55 m were made before excavating to the foundation level in such a way as they may function as struts against the tuff walls, particularly for module F, at an elevation of approximately +7.50 m (see Figure 8). This was dictated by the need to counter a barbican whose foundation is presumed at an elevation of +3.00 – an elevation where the tuffs settle, as confirmed by the latest probes on November 28, 2006.

As regards the procedures for building these floors, as already discussed, by exploiting the deep foundation type, already at an elevation of +3.20 m pile-piers were built, constituting the vertical support of the deck at that elevation.

This is followed by the reinforcement and the casting of the setpa which support the floors at an elevation of +6.60 m and +9.65 m. Having completed these two phases, after appropriate curing of the castings, the excavation took place by indenting the existing tuff bank, which was further lowered to the installation elevation of the connection slab in the foundation, to be followed by the set section excavation to build the plinths.

As may be easily intuited, these cases considerably slowed the working operations; it must not in fact be forgotten that the need to make the decks to an elevation of 9.65 before reaching foundation level led to excavations under cover, with the use of appropriate equipment. Additional provisional works (plating and mini-piles) were needed to stabilize the masonry septa nearest to module I, even if not provided for in the design phase.

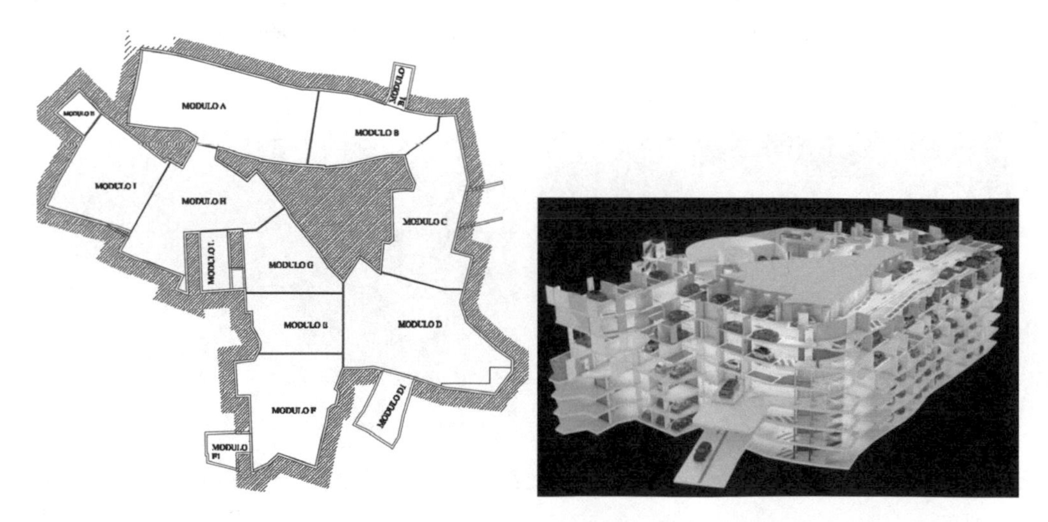

Figure 8. Layout of the modules of the parking and a 3D rendering.

Figure 9. Pictures of the construction of structures.

Figure 10. Old vehicles found during the works on the left, the parking today on the right.

9 CONCLUSIONS

Underground car parks in urban centers are a functional solution to chronic traffic and parking problems affecting large cities.

The Morelli car park, even if it represents a challenge from a technical and executive point of view, has allowed an optimal and functional reuse of one of the abandoned caves in the subsoil of Naples, guaranteeing a service to the citizens and at the same time giving them a unique scenario, today is a venue of events and guided tours.

REFERENCES

Evangelista, A., Aversa, S. 1994. Experimental evidence af non-linear and creep behavior of pyroclastic rocks. *Proc. Course on Visco-plastic behaviour of geomaterials.*

Aversa, S. 1989. Comportamento del Tufo Giallo Napoletano ad elevate temperature. *Tesi di dottorato di Ricerca in Ingegneria Geotecnica, Università di Napoli.*

Evangelista, A. 1980. Influenza del contenuto d'acqua sul comportamento del tufo giallo napoletano. *Atti del XIV Convegno Nazionale di Geotecnica, Firenze.*

Bringiotti, M. 2003. *Guida al Tunneling 2nd ed.* PEI.

Bringiotti, M., and Bottero, D. 1999. *Consolidamenti e fondazioni 2nd ed.* PEI.

Tunnels and Underground Cities: Engineering and Innovation meet Archaeology, Architecture and Art, Volume 1: Archeology, Architecture and Art in underground construction – Peila, Viggiani & Celestino (Eds)

Moncenisio, from Myth to history TELT – Tunnel Euralpin Lyon Turin and the collection of historic engravings on the Frejus tunnel

M. Virano, G. Dati, M. Ricci & G. Avataneo
TELT-SAS, Turin, Italy

ABSTRACT: TELT possesses a collection of original documents concerning the original Fréjus Rail Tunnel. By using them as a guideline, this paper reconstructs the history of the Fréjus tunnel, showing that it has plenty of lesson to teach us. The elements presented include: Cavour's closing statement during the parliamentary debate; The affair of the Fell Railroad, which warns us that new technologies have often an unforeseeable impact; The digging of the tunnel, whose completion was made possible by the development of the pneumatic perforator; The follow-on impact of the tunnel, which led to the digging of several other transalpine conections and which caught the eye of the public, leading to the construction of the so-called "Bogorama". As TELT digs the new Mount Cenis Base Tunnel, we can look back at the history of the first tunnel to cross the Alps as both an example and as a learning experience.

1 THE TELT HISTORICAL COLLECTION AND THE SAGA OF THE FRÉJUS

To pass on the archaeology of underground excavations, collecting its traces and placing them at the disposal of enthusiasts and scholars alike. This is the duty that TELT – Tunnel Euralpin Lyon Turin, feels called up to accomplish. It has led TELT to put together a collection of engravings, dating back to the second half of the 19th Century, which reconstruct the history of the Fréjus Tunnel.

This collection, which TELT saved from dispersal, includes prints, both woodcuttings and original photos, taken from Italian, French and English newspapers of the time. The material is displayed in an exhibition room in TELT's Turin seat. There are five thematic sections: The Time before the Tunnel, the Fell Railroad, Excavation and Construction of the Tunnel, The Operating Line, Inauguration and Celebrations.

The collection is a mine of historical and technical information. It comprises 60 woodcuttings, 2 original photos, and a four-and-a-half-metres-long panel which displays the track of the railroad between Bussoleno and Modane. Furthermore, it also includes a series of original documents and books that make it possible to reconstruct the main political and technological developments of the excavation, such as the project for Germano Sommeiller's "perforating machine" and a table with the longitudinal sections of the tunnel.

The TELT offices, which are located in the Officine ferroviarie Grandi Riparazioni di Torino complex, have been designed as a Speaking Environment. The structure of the set-up and the use of infographics and images allow the visitor to discover many aspects of the Turin-Lyon line: from the history of the project to the geological composition of the Alps, from the characteristics of the Mont Cenis Base Tunnel to the excavation techniques used, to the connections with the TEN-T Network and the New Silk Road.

Figure 1. A portion of the exhibition room in TELT's Turin headquarters.

2 THE GREATEST OF ALL MODERN ENDEAVOURS

The Fréjus Tunnel was started in 1857 by the Kingdom of Sardinia, but it was the unified Italian state that inaugurated it on September 17, 1871, almost a year to the day after the taking of Rome. The original aim was improve communication between two territories belonging to the same state, but the transfer of the Savoy region to France in 1860 made the tunnel a cross-border connection instead. The excavation began with the miners painstakingly digging by hand the holes for the mining charges, but it was accomplished thanks to a massive use of mechanical drills. These examples offer a glimpse of the scope of the changes that involved the Fréjus Tunnel. Its excavation represented a massive engineering and financial challenge and its completion is a testament to the qualities of the Sardinian government of the time.

2.1 *A Bold Decision*

The first of the qualities demonstrated by the Savoy state was efficiency. Its Chamber approved the project for the Fréjus Tunnel on June 27, 1857 and the Senate gave its approval two days later, with a forecasted expense of 41.1 million liras. King Victor Emmanuel II signed the relative law on August 15 and digging began on August 31. Overall, once Parliament approved the project, it took little more than two months to start it, and the King was at work signing laws even during the holiday of Ferragosto. Any comparison with modern times can only be unfavourable to current proceedings, and this comparison becomes even more lopsided if one looks at the content of Cavour's closing argument during the debate at the Chamber.

In his opening statements, the Savoy Prime Minister had gone over the economic benefits that the completion of the tunnel was expected to produce, such as an increase in Piedmont's rice exports and the deviation towards Genoa of the trade with Genève currently going through Marseilles. MP Menabrea had then highlighted the geopolitical implications of establishing a direct rail connection with France. Once the Suez Canal, the digging of which would begin two years later in 1859, opened, the port of Brindisi could then become the European terminal of a commercial line towards India and China. To this respect, the TELT collection includes a poster with the 1891 winter schedule of the trans-European railway line that was indeed established after the opening of the Fréjus Tunnel. As the schedule shows, this line went from London to Brindisi and then Palermo, and it connected Paris with Rome in just thirty hours. This vision of using the ports of Southern Italy as terminals of trade with the East is far from being outmoded. To the contrary, it has gained prominence again in recent years, following the unveiling by Chinese president Xi of the project for establishing a new Silk Road, the so-called "Belt and Road Initiative", to strengthen trade links between Asia and Europe.

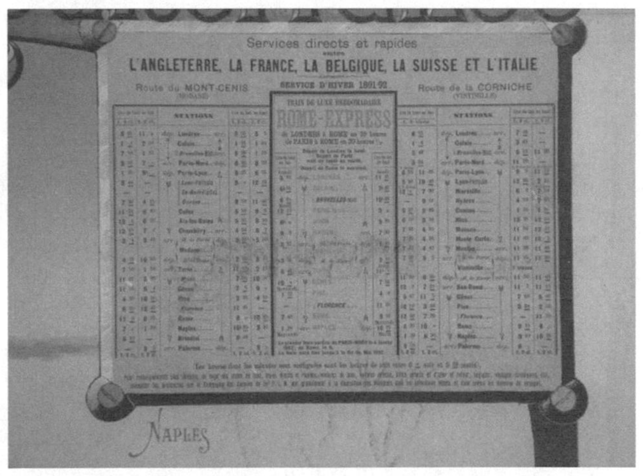

Figure 2. The illustrated schedule of the London to Palermo line.

In his concluding remarks on the project, however, Cavour set aside all practical considerations in favour of a declaration of principle. It took the form of a rebuttal to the statement of MP Moia, his main opponent. Moia had declared himself not opposed to the project, but he had also highlighted its technical and financial risks and had suggested proceeding with the maximum possible level of caution. To this, Cavour replied that a policy based on half-hearted measures and on repeat studies commissioned only to postpone and delay decisions, a policy, therefore, which was not willing to assume responsibility for its choices, would end up blocking all initiatives, with disastrous consequences for the whole country. It is worth repeating the actual words of Cavour, because they have kept intact all their value in today's world:

> *"Gentlemen, the endeavour that we propose here, it's not worth denying it, is a gigantic endeavour, its completion will however bring glory and benefit to the country. We have never hidden to you that we are convinced that this endeavour cannot be accomplished without overcoming huge, immense difficulties; we have never dissembled the scope of the responsibility, so much so that we are here asking for a confidence vote. But, if the difficulties to face are many, no less is the hope that we are capable of overcoming them.*
>
> *Great endeavours are not accomplished, however, unless one condition is met, which is that those who are tasked with bringing these tasks to completion have a solid, absolute faith in their success. If this faith is not present, we must not attempt great things, neither in politics, nor in industry; if we did not have this faith, we would not have come here today to insist before you, placing on our head such a heavy burden. If we were hesitant men, if we let ourselves be scared by the thought of responsibility, we could adopt the system proposed by MP Moia, even tough, in the end, it could result fatal. As we are, however, unaccustomed to half measures, not used to proposing a timid, vacillating and perplexed politics, we could not accept his proposal, and we invite you to weight on your scales the only two rational systems: that of the execution, via a contract immediately established with the Laffitte Company, or of the deferment to other times of this brave attempt. [...]*
>
> *"I am convinced, gentlemen, that you will share our confidence. I hope that you will express a clear vote. If you share our faith, I call you to vote resolutely with us; [...] if a doubt torments you that in the bowels of the mountain which we wish to cut open are hidden all manners of difficulties, obstacles and dangers, I call you to reject this law, but do not condemn us to adopt a middle line, which in this case would be fatal. [...] I trust that you will always follow an open, decisive politics. If you adopted today the Moia proposal, you*

would inaugurate a new system, and I would lament it extremely, not just because we would lose this magnificent work, but because such an act would sign the death knell for the future political system that Parliament will be called on to follow. We had the choice of the way; we have preferred that of resolve and daring; we cannot remain in the middle; it is for us a vital condition, an unavoidable alternative: to progress, or to die.

I hold firm faith that you will crown your work with the greatest of all modern endeavours, by deliberating the piercing of the Mont Cenis."

This speech is a clear example of Cavour's political vision and it shows all his foresight in identifying the dangers of a political body that refuses to make decisions, which is one of the woes of our times. Along with efficiency, far-sightedness is indeed the most important quality shown by Savoy governments concerning the Mont Cenis Tunnel. It was not confined just to Cavour, as the project started long before he came to power. This quality was well expressed by MP Luigi Federico Menabrea's declaration of vote:

"I believe in the Suez canal opening in a foreseeable future, because I am sure that Europe will come to understand that to survive, it is necessary to open up towards the indies and Chinese sea. This will balance the power of an opponent nation growing up really quickly to become a giant beyond the Atlantic. I am saying that the future of our country is ensured, that we will reach a prosperity level never seen before, because it will be the route for big part of transit and trade between Europe and Orient".

2.2 *A Forward-looking Vision*

It was in the year 1844 that Belgian engineer Henry Maus was given the task of evaluating the possibility of establishing a transalpine rail connection between Savoy and Piedmont. At the time, there were only a few kilometres of railway lines in operation throughout the whole Italian Peninsula, and none of them were in the Kingdom of Sardinia. Nevertheless, King Carlo Alberto and his government were already aware of the importance that transalpine rail connections would have one day. Not just to bind together the domains of the Crown, but especially in order to place Piedmont solidly within the growing network of cross-European trade routes. The latter is a point that Cavour would make again during the parliamentary debate of 1857. The TELT Collection includes a copy of the final study presented by Maus in 1850. It also includes a rather

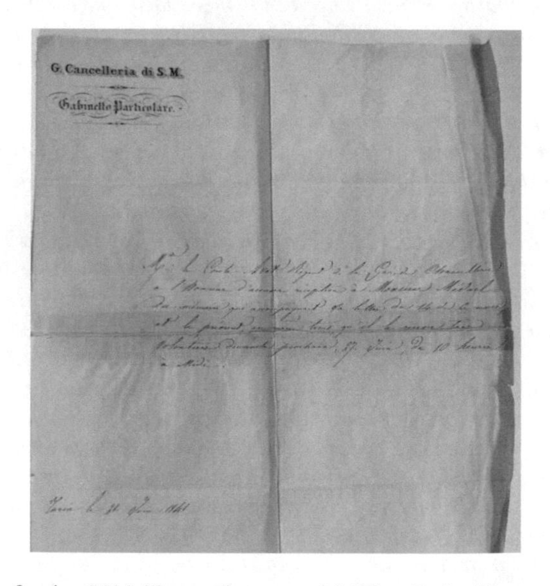
G. Cancelleria di S.M.

Gabinetto Particolare.

Figure 3. The letter confirming Médail's appointment with King Carlo Alberto.

curious document that shows the king's personal interest in the matter. It is an autographed letter by Count Avet, head of Carlo Alberto's chancellery, confirming the appointment with the king to Joseph-François Médail, a lowborn customs official. The appointment was to take place "Sunday, 27th of June [1841], between 10 a.m. and midday". It is a day, an hour, and a visitor's rank that are quite unusual for an audience with the king. The subject? Discussing the "Project of excavating the Alps between Bardonecchia and Modane" that Médail had sent to the king and his government the previous 20th of June.

3 CROSSING THE MONT CERNIS

Médail's project did not lead to any actual initiative. It was too far ahead of its time, considering that the first operational railway in the Kingdom of Sardinia opened only in 1848, seven years after Médail had presented his project. This first line was the 8-km-long track between Turin and Moncalieri, itself just the first portion of the Turin-Genoa line. The whole line opened only in 1853, having to wait the completion, after an excavation entirely by hand, of the Giovi Tunnel. Meanwhile, people kept crossing the Mont Cenis in the same way they had followed in the past. The TELT Collection can help us to reconstruct how that took place, as it includes a series of original woodcuts from the middle of the 1860s, a period before the opening of the tunnel.

For centuries, the only way to cross the Mont Cenis was a narrow mule trail, sporting 33 hairpin turns on the Savoy side and 77 on the Piedmont side. The first major improvement was the construction, between 1803 and 1809, of an actual, 37-km-long road. The road was commissioned by Napoleon, who wished to be able to move his armies quickly between France and Italy. Napoleon, after all, had already lead two major campaigns in Northern Italy, in 1796 and 1800, and since 1805 he held, along with the crown of Emperor of France, also that of King of Italy. Furthermore, the renewal of hostilities with Austria in 1804 had increased the importance of the Italian theatre and the consequent need to possess a fast way to cross the Alps. The opening of the road allowed the Mont Cenis to be crossed by coach, drastically reducing both the travel time and the risks of the journey.

Woodcuts from the TELT collection show us the developments that followed the opening of the road. With the arrival of railways to both sides of the mountain, stopping at Saint-Michel-de-Maurienne and Susa respectively, by the mid-1800s it remained only the pass itself that had to be crossed by traditional means. At the time, travel between Paris and Turin took 35 hours, with three legs by rail interrupted by one by boat along the Lac du Bourget and one by coach to cross the Mont Cernis, with the latter including parts via sled in winter. Even with such a complex system, the arrival of the railway promoted a rapid increase in exchanges. During the year 1863 alone, the pass was crossed by over 40,000 travellers and 22,000 metric tons of cargo.

Figure 4. The crossing of the Mount Cenis in winter according to an 1865 French woodblock print.

4 THE FELL RAILROAD

The last main innovation before the opening of the tunnel was the so-called Fell Railroad, whose history is also told by the TELT Collection. This line, inaugurated in 1868, ran parallel to the Napoleonic road. In order to be able cross the steepest slopes, it utilised a three-rail system, a predecessor to the modern rack railway, patented by British engineer John Barraclaugh Fell. The Fell Railway is important for a series of reasons. The first is eminently practical: while it operated, it cut down the time needed to cross the Mont Cenis from sixteen to just four hours. Its importance, however, is mainly due to its economic and industrial relevance. The construction of the line was proposed in 1861 by a group of English investors that included Fell himself. At the time, the Fréjus excavation had already been going on for four years. In fact, the line was designed as a temporary measure, with the idea that it would cease its operations once the tunnel was finished. In the estimates of Mr Fell and his partners, completion of the Mont Cenis tunnel would take enough time to allow their line to turn a profit in the meantime.

Figure 5. The Fell railroad in operation. Clearly visible is the third track specific of the Fell system.

A series of delays, however, prevented the Fell Railroad from opening before 1868, while the resolution of the problems that had plagued the Sommeiller pneumatic system allowed the conclusion of the excavation of the Fréjus Tunnel by mid-1871. Therefore, the Fell Railroad was forced to close down after just three years of activity. While it had, during its operation, transported over 100,000 passengers, the result was a net loss for its investors. This story is an example of the sometimes unforeseeable consequences produced by technological change. It can also serve as a warning, which is still valid today, on the need to evaluate carefully those projections that go beyond the short term. Predictions, as the saying goes, are always difficult to make, especially when they concern the future. We can also add that they become increasingly uncertain the more you go beyond the present. Especially, as was the case for Fell, if you are dealing with technologies that have not yet had the time to demonstrate their full disruptive potential. We will come back again to this point in the course of this paper.

5 DOUBTS ABOUT THE TUNNEL

In defence of Fell, there were actually several reasons to believe that the excavation of the Fréjus Tunnel would have taken a very long time, and even to doubt its eventual completion. It was the first attempt to dig a tunnel through the Alps. The only applicable reference was the Giovi tunnel, dug as part of the construction of the Turin-Genoa line, which ran through the much smaller Apennines. This tunnel was three kilometres long and it ran at a height of 360 metres over sea level. It had taken eight years to dig and, when it opened in 1853, it had been the longest tunnel ever completed. The proposed Fréjus tunnel was over 12 kilometres long, four times the Giovi, and it had to be dug through a tall mountain, where information on the types of soil to face was very scarce. Furthermore, there were many doubts concerning the ability of the buyer, i.e. the Kingdom of Sardinia, to finance the project. Those were the same years of the Risorgimento, the process of unification of Italy. At the beginning of the dig, the Savoy coffers showed a debt of 680 million liras. This was due to the First War of Independence of 1848 and to the Sardinian participation to the Crimean War in 1854–56. The start of the Second War of Independence then left only 500,000 liras available to the Fréjus project for the whole year of 1859, further delaying the works.

After considering the technical and financial challenges facing the Fréjus project, it can be understood why Fell was by far not the only one who had doubts about the completion of the tunnel. Another sceptic was the French state. When France acquired the Savoy region, it accepted with an accord signed in 1862 to finance a portion of the endeavour. The accord stipulated that France was to pay 19 million liras, but only if the tunnel was completed within the following twenty-five (!) years. It also included an incentive mechanism according to which, if the tunnel opened before the deadline of 1887, the French would pay an additional 500,000 liras for each year saved, or 600,000 for each year before 1877. Since the tunnel was completed already in 1871, France ended up paying 26,100,000 liras, which corresponds to more than a third of the total cost of the dig, estimated at 70 million by an 1873 study. It is worth recalling that the law of 1857 that established the project had foreseen an expense of 41.1 million. Cost overruns for big projects were, indeed, as common at the time as they are today. The Suez Canal, which was dug at the same time as the Fréjus tunnel, ended up costing more than twice what was originally envisaged.

6 THE SOMMEILLER SYSTEM

Completion of the Fréjus Tunnel in just fourteen years, long before the estimates of Fell and of the French, was made possible by an important technological innovation: the Sommeiller pneumatic perforator. Up until that moment, the traditional digging method was for the miners to prepare by hand, one by one, the holes where to place the explosive charges. Following this method, during the first year, digging through relatively easy rock, the excavation advanced by about four hundred metres. Such a speed would have required thirty years to dig through the more than twelve kilometres of the tunnel. During the preparatory phase of the project, a perforating machine made by the British inventor Bartlett had been tested as an alternative. This machine used steam to move a mechanical drill and it produced good results during the essays. However, the problems connected with safely bringing steam power to a machine destined to be placed up to six kilometres inside a mountain could not be easily solved. A government committee then tested in Genoa in 1856 a modified version of the Bartlett machine, which used compressed air in conjunction with steam, but this version was judged too complex and difficult to handle. It was Savoyard engineer Sommeiller who found a solution. He developed a simplified version of the Bartlett perforator that used only compressed air. This innovation permitted to solve two key problems at the same time. First of all, the possibility of carrying the compressed air via a network of tubes allowed the power generators, in this case the compressors, to be kept in fixed buildings placed outside the galleries. These compressors could then be powered very cheaply using the hydraulic power of mountain torrents. Furthermore, the ample availability of air allowed to safely remove the fumes generated by the explosion of the mining charges by forcing the atmospheric circulation within the galleries.

Figure 6. An early model of the Sommeiller Perforator (left) and its final version (right).

The first years, as it is the case with all innovative technologies were not easy for the Sommeiller perforator. The images in the TELT collection show us the evolution of the models used during the excavation. The original compressors proved prone to failures and they had to be re-designed completely. Even more difficult was finding a correct design for the drill heads, which tended to break down or wear out very fast. In the sole year of 1862, the workers ran through more than 72,500 different drill heads. The best shape, it was found out, adopted a point squared like a "z". These and other teething problems were faced and gradually solved one after another and the digging speed progressed accordingly year after year until, by 1870, it was four times that of the original manual dig.

Besides allowing the tunnel to be completed much faster than it would otherwise have been possible, the use of compressed air in the Fréjus tunnel had many other repercussions. Some are easily imagined, such as the creation of a new whole sector, that of pneumatic excavation. Indeed, Belgium started a feasibility study concerning the use of pneumatic perforators in its coalmines already in 1863, when the Sommeiller machines were just beginning to show their worth. Other repercussions, and with this we come back to the subject of the unforeseen disruptive effect of new technologies, were less obvious. The most important is probably the one that concerned the rail sector: it was by reading an article on the machines used in the excavation of the Fréjus Tunnel that American inventor George Westinghouse got the inspiration of using compressed air to power a distributed breaking system affecting all wagons of a train. This system is, with all the improvements brought by more than a century of use, the one that we still use today.

7 THE IMPACT OF THE FRÉJUS TUNNEL

7.1 *The First of its Kind*

The first, and most important, effect of the completion of the Fréjus Tunnel was to demonstrate that it was actually possible to dig galleries through the Alps. This led to the excavation of several additional transalpine tunnels in the following years. Indeed, works for the Gotthard Tunnel started already in 1872, just one year after the opening of the Fréjus. Furthermore, the story of the project showed that it was possible to accomplish these works with a minimal loss of human life. During the thirteen years that it took to dig through the Fréjus, the construction suffered a total of 48 deaths. It is for sure quite a high number, but it is still low if compared, for example, with the 177 official deaths during the ten years of the Gotthard dig, which in addition do not include deaths for sickness or malnutrition. Furthermore, of the 48 Fréjus deaths, eight were due to fights, one was a suicide and four were due to the explosion, probably caused on purpose, of the Fourneaux powder magazine. The main killer, however, was the epidemic of cholera that developed on the Italian side in 1865, which killed 18 diggers and caused further 60 deaths in the nearby city of Bardonecchia. The deaths directly liked to work site accidents were therefore, on average, less than two per year, which is, no matter how one looks at it, a very low number, especially if one considers the technological level of the time. Such a reduction in the total amount of deaths constitutes an undeniable merit of the management of the Fréjus excavation.

7.2 *The Bogorama*

The impact of the Fréjus Tunnel was not limited to technical circles. Just like the other great works of the century, it attracted the interest of the press, of artists, and of the general public as well. At the beginning of the year 1870, the two halves of the tunnel were still separated by a kilometre and a half of rock, but the dig was by that time advancing rapidly and the people of nearby Turin started talking about the forthcoming inauguration of the tunnel. After all, such ideas were part of the spirit of the time, as the main news then was the completion of another great work that had a big impact on Italian commerce: the Suez Canal. Even the great Italian opera writer Giuseppe Verdi had been asked to contribute to the inauguration ceremony of the Canal, which took place with great pomp on November 17, 1869. Verdi wrote for the occasion the *Aida*, even though some technical delays eventually forced its replacement with the *Rigoletto*. It should not surprise therefore that the free spirits of Turin's Artists' Society took advantage of the 1870 carnival to build in Piazza Castello a stand that was a tribute to the two inaugurations of the year: one, Suez, which had just happened, and the other, the Fréjus, which was expected to happen shortly. The structure, of which the TELT collection includes an original albumen print photo, was a gigantic pharaoh's head, whose mouth gave access to a pavilion containing a single, 120-metres-long painting. The painting represented the vistas encountered during an imaginary voyage from Bardonecchia to Suez. The front of the mask showed the writing "Bardoneccio – Suez – Bogorama", referring to the two ends of the voyage and to the "Great Bogo" goliardic- chivalric order, founded years before by the same members of the Artists' Society. The Bogorama, as the whole work was called, had a large success with the public and it remained in Piazza Castello until Lent, but it then suffered a sad ending. It was bought by a Frenchman who wanted to replicate in Paris the success that the Bogorama had incurred in Turin, but it was destroyed instead in one of the fires caused by the constitution and then suppression of the 1871 Commune. The photo in the TELT collection is therefore the only picture we have left of the Bogorama, which in turn is a signal of the interest with which the public followed the progress of the Fréjus Tunnel.

Figure 7. The Bogorama.

8 CONCLUSIONS

The Fréjus Tunnel was the first alpine tunnel ever dug. The Mont Cenis Base Tunnel that TELT is digging 800 metres below it comes, instead, last among the various alpine base tunnels. Because of this, however, it can benefit from the best practices, both technical and managerial, developed during the construction of the other seven next-generation alpine tunnels.

This is not, however, enough. The ambition of TELT is to be a pathfinder in employing cutting-edge technologies, just as the technicians and engineers who were protagonist of the Fréjus endeavour did during their time. It is a path that the Franco-Italian company is following through high-profile technical and scientific partnerships with institutions such as the Turin Polytechnic and the University of Bologna in Italy or the Ecole Nationale des Ponts et des Chaussées and the Centre Etudes Tunnel Universitaires in France. TELT also aims to set the highest standards for governance, with the benchmarking of great works and by applying, for the first time in Europe, the anti-Mafia legislation to the whole of the project, independently from the nationality of the individual sites.

REFERENCES

Antonetto, R. 2001. *Fréjus. Memorie di un monumento*. Turin: Umberto Allemandi & C.

Bruno, A. (ed.) 2012. *George Westinghouse. Un genio del XIX Secolo*. Turin: Circolo George Westinghouse.

Ceresa, F. (ed.) 2013. *Bardonecchia, la nascita di nuovo borgo*. Turin: Politecnico.

Figuier, L. 1884. *Les nouvelles conquêtes de la science. Grand tunnels et railways métropolitains*. Paris: E. Girard & A. Boitte Éditeurs.

Figuier, L. 1884. *Les nouvelles conquêtes de la science. Les voies ferrées dans les deux mondes*. Paris: E. Girard & A. Boitte Éditeurs.

Tunnels and Underground Cities: Engineering and Innovation meet Archaeology, Architecture and Art, Volume 1: Archeology, Architecture and Art in underground construction – Peila, Viggiani & Celestino (Eds)
 ISBN 978-0-367-46574-2

Interdisciplinary research in geotechnical engineering and geoarchaeology – a London case study

F.K. Vonstad, P. Ferreira & D. Sully
University College London (UCL), London, UK

ABSTRACT: Collaboration and integration are vital project functions enabling a development project to achieve completion on time and on budget. The development of interdisciplinary approaches to research is considered a powerful mechanism that is vital to facilitate more efficient work practices in construction and in research. There is little empirical evidence currently suggesting how this collaboration can be achieved directly on sites within large infrastructure projects. By examining surrounding geotechnical borehole scan logs and archaeological fieldwork logs of a site, this research assesses the loss and importance of geoarchaeological information within a London Case Study, and how the geotechnical results found show a need for interdisciplinary action. Neighboring case studies indicates that there is potential for significant finds and deposits to survive un-truncated on the Case Study site, which is in proximity to an Archaeology Priority Zone. In addition to showing the potential loss of geoarchaeological information, the research shows how an interdisciplinary approach to geotechnical engineering and geoarchaeology on developments can save time, money and reduce risk for surprise archaeology on sites.

1 INTRODUCTION

It is commonplace for major cities to have historical significance as population centers, and therefore extensive geoarchaeological information stored in the subsurface layers. Geological archaeology, also known as geoarchaeology, and geotechnical engineering are the main subsurface investigators in development projects. Both disciplines perform analysis which are knowledge-intensive tasks (Stein, 1986:505), and as found by Goldberg and Machphail (2006:29) geoarchaeology, like geology and geotechnical engineering, is a field-based endeavor that relies on empirical data accumulated from a site (Caputo, 2003:688). As construction projects teams grow in size and complexity, research and cross-disciplinary dependences become more copious, varied and complicated thus increasing the need for collaboration to ensure progress (Faraj and Xiao, 2006:1155, Jørgensen and Emmitt, 2009:225). Interdisciplinary approaches to below-ground research based on synergistic work practices is becoming an increasingly popular subject within academia and industry, as such collaborations are capable of providing a wider scope of solutions and information about problems encountered. Research in this area has been requested by numerous researchers (Kruse, 2006:17, D'agostino & Tocco, 2013:289, Viggiani, 2013:3), as well as development authorities (City of London, 2012:16) and heritage conservation bodies (Historic England, 2016:8). There are many occurrences of sites where the archaeological research is limited, from constrictions such as costs, time and access, where the geoarchaeological investigation could have been performed as an interdisciplinary collaboration with geotechnical engineering investigations. The anonymous

Case Study described in this paper investigates the loss of geoarchaeological information at a development site in Bermondsey, London, due to a lack of borehole research for archaeological purposes and the lack of collaboration with the engineers on the geotechnical boreholes performed. The paper investigates the potential geoarchaeology on the site, based on archaeological fieldwork undertaken in the area and geotechnical core samples taken in proximity to the site. For archaeology it is crucial to be more integrated in the development process to survive as discipline. Almost 90% of archaeological material comes from development sites, and with recent policy changes in the UK, less archaeological investigations are required by developers. By finding more collaborative and synergetic work practices, archaeology could be performed in a more efficient and cost and time preserving manner, as well as reducing the risk of surprise archaeology that could delay development projects indefinitely, through earlier detection and mitigation in the planning process.

Research Questions:

1. *Can the surrounding archaeological and geotechnical fieldwork and borehole logs provide information for the Case Study?*
2. *What potential buried geoarchaeological information was lost in the Case Study, based on surrounding archaeological and geotechnical data?*
3. *Can geotechnical boreholes provide information for geoarchaeological research?*

2 CASE STUDY

The Case Study Site lies just to the east of an Archaeological Priority Zone (APZ), as designated in Southwark Council's Unitary Development Plan. The Case Study lies 1.7km to the south of the River Thames, to the south-east of an area of relatively high natural ground known as the Bermondsey Eyot. Geological information for the site and its vicinity suggest that it would have been situated on a slight topographic high-point and may therefore have been utilised by human groups seeking to exploit the wetland resources of the Thames from the Late Upper Palaeolithic period onwards. Approximately 350m to the North-West of the site a large glacial lake existed in Prehistory. The geology of the area consists of Pleistocene River Terrace Gravels overlain by Holocene Fluvial sediments (Sidell et al, 2000:2). Prehistoric artefacts and evidence of occupation have been discovered on various sites in Bermondsey as seen by Ridgeway (Ridgeway, 1999:74) and by Cotton and Greene (Cotton & Green, 2004:134). Finds in the area has included flint tools and some evidence of metalwork. The desk-based evaluation has shown that there is potential for survival of ancient ground surfaces (horizontal archaeological stratification). There is also potential for survival of archaeological cut features and other organic artefacts. However, such survival is likely to be limited in certain areas because of existing basements and previous building work (Rogers, 1990:230, Mackinder, 2000:6, Knight, 2002:5).

The Case Study in Bermondsey originally had a full archaeological investigation planned for the site, based on the high risk of finds, particularly environmental ones. Due to delays in the project, and the development company being changed to one less inclined for archaeological investigations, only minor trenching was undertaken on the site. In the original development plans for the historic environment, 8 boreholes had been planned in various areas of the site. No boreholes were taken besides the early geotechnical boreholes, which did not search for any geoarchaeological parameters at the time. The Case Study has several sites where finds of archaeological importance were discovered in its direct vicinity, and is in an area where ancient landscapes and settlements have been known to be found. There is no published archaeological record from the commercial trenching that was done, and no published geotechnical borehole scan logs.

3 METHODOLOGY

The Methodology for the project was challenging, as not much archaeological data was gathered from the site as a whole, making data collection from surrounding fieldwork and geotechnical logs the only logical place to look for further information. The project is based on using the surrounding data to create an image of how the site fits in the wider geological and human history of the area. This in no way can replicate what could originally be found on the site, but is helpful for the archaeological understanding of the site in relation to human occupation in the Southwark/Bermondsey area.

3.1 *Data Collection Plan*

An inductive qualitative research methodology was adopted for the research purposes of this paper, based on the work by Eisenhardt and Graebner (2007:25) and Glaser & Straus (1973:45). The inductive qualitative research approach was chosen over other methods for its emphasis on the rich, real world context in which phenomena occur (Eisenhardt and Graebner, 2007:25), and how the initial decisions for theoretical collection of data are based only on a general problem area, only beginning with a gross concept of structure and process, so that the theory can be conceptualized and formulated as it emerges from the data. This reflects the way data was collected and the how it was utilized to create a hypothesis in this paper. Data was gathered from 4 main sources for the research case study:

1. https://www.archaeologydataservice.ac.uk
ADS gathers archaeological data and reports in the UK.

2. https://www.mola.org.uk/research-community/resource-library
MOLA is the lead archaeological contractor in London with detailed field reports.

3. http://www.bgs.ac.uk/data/boreholescans/home.html
BGS is the official geotechnical data source for borehole logs in the UK.

4. Unpublished evaluations and background studies from the development company

Following Eisenhardt and Graebner (2007), a purposive sampling strategy was implemented for data gathering from the 4 sources. Fieldwork logs from ADS and MOLA where gathered from an archaeological perspective of the area, and geotechnical borehole log from BGS were gathered following set criteria to ensure the information richness and relevance of the data and to more easily observe patterns:

1. Proximity to the Case Study site
2. Descriptions of potential archaeological "inclusions" (Geotechnical Logs)
3. Descriptions of archaeological finds and conditions (Archaeological Logs)

Additional information was gathered from the desk-based study created from the project, the planned works, the edited plan and the final assessment of the site. All of these papers remain unpublished and due to the requested anonymity of the case study, the material will only be referred to as development papers and plans.

3.2 *Data Analysis Plan*

Data analysis for the research is based on a qualitative inductive approach. Qualitative data analysis is simply the process of examining qualitative data to derive an explanation for a specific phenomenon. Qualitative data analysis gives an understanding of the research objective by revealing patterns and themes in the data. In this case, the research questions were used as a guide for grouping the datasets, and analyzing the raw data gained. The following steps were taken in the analysis of the data:

1. Sift through data for relevant archaeological and geotechnical logs
2. Sort by location, depth, archaeological finds and soils encountered
3. Pattern identification based on the manual inspection of the logs to identify emergent themes from which new knowledge could be inducted

4 FINDINGS

The research was based on the relevant archaeological fieldwork and geotechnical borehole logs, in addition to the desk-based assessment and plans made by the development company for the site. A total of 167 geotechnical borehole logs in the area were considered for the gathered datasets, and a total of 128 archaeological logs. From the total 295 considered borehole logs and fieldwork studies, 9 examples from each discipline were chosen based on the criteria mentioned in the Methodology.

The total 18 chosen fieldwork studies and borehole logs, here from referred to as data sets, were systematically examined to extract information for a comparative assessment to the Case Study in focus. Based on the summarized information interpreted from the data sets, a suggestion of the potentially lost geoarchaeological information was made. In Figure 1 below the spatial spread of the case studies selected:

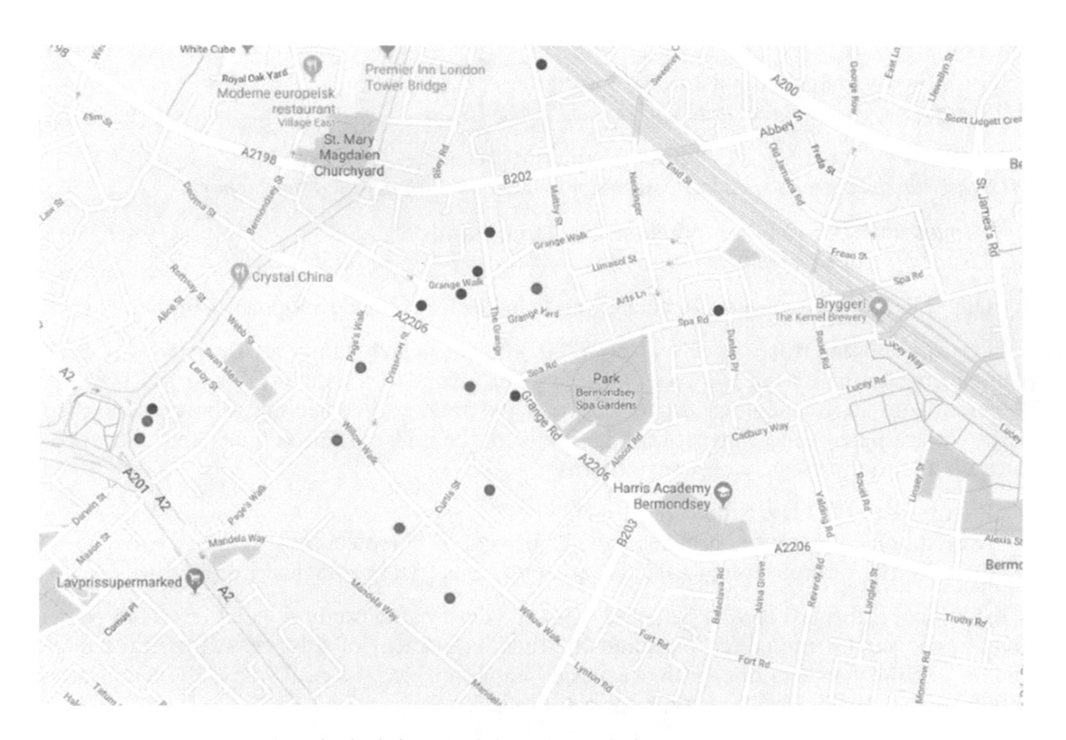

Figure 1. Blue dots: Archaeological sites, Red dots: Geotechnical sites.

The analysis and evaluation of the datasets has shown that there is potential for survival of ancient ground surfaces (horizontal archaeological stratification) around the site, as well as survival of several organic artefacts. Finds from the archaeological datasets where collated in Table 1 – Archaeological Finds, which is shown below. The datasets shows Case-numbers which can be referred to in the MOLA archaeological records and the ADS, for repeatability.

Table 1. Archaeological Log Finds.

Case	Depth	Organic	Shell	Flint	Pottery	Other
MLO58452	c. 0.76m	N/A	N/A	Neolithic	1–2nd C.	Cess Pit
FLO18116	N/A	Bone Hair Pin	N/A	Mesolithic	Roman	N/A
MLO073323	N/A	N/A	N/A	N/A	Neolithic	Prehistoric Pit
MLO105026	N/A	Roman Farm	N/A	N/A	N/A	N/A
MLO23838/ ELO2701	N/A	Land Surface - Prehistoric	N/A	N/A	N/A	Leather Shoes – Roman/Saxon
MLO63702	N/A	Peat – Bronze Age	N/A	Axes, Tools - Neolithic	N/A	N/A
TW53356-179230	c. 2m	Charcoal	N/A	N/A	N/A	CBM, Pipe Stem
MLO60131	c. 0.40m	Animal Bone	N/A	N/A	Medieval	Pipe stems, ceramic drains
MBZ06	c. 2m	Peat, log with axe cuts (prehistoric), charcoal, animal bone	N/A	N/A	Medieval	CBM, Glazed Tile, Horn Core

Based on the case studies the potential loss of geoarchaeology included:

1. Prehistoric landscape information
2. Information regarding land use
3. Organic artefacts of material such as leather and bone, and organic paleo-environmental proxy indicators
4. Information on settlement movement and landscape changes

In summary:

1. 7 of 9 case studies had organic finds
2. 4 out of 9 case studies organic finds were related to landscape
3. 3 out of 9 case studies had Neolithic finds
4. 9 out of 9 case studies had archaeological material

The prehistoric land surface recorded at the neighbouring archaeological site Alaska Works (*MLO23838/ELO2701*) to the north-east of the Case Study indicates that there is potential for finds and deposits from this period to survive un-truncated on the gravels of the site. In addition, the gravels in the south of the site may be overlain by organic deposits (as observed to the south of the site) with the potential to preserve cultural material and palaeo-environmental proxy indicators (pollen, plant macros, microfauna). The environs of the site would have offered favorable conditions for hunter-forager communities during the Upper Paleolithic and Mesolithic (c 10,000–4,000 BC) periods, as is attested to by finds from various surrounding sites and finds listed above, such as Bricklayers Arms Goods Depot (*MLO63702*), which provides important information about the Neolithic use of the area.

The geotechnical borehole scan logs were surprisingly informative from a geoarchaeological perspective, as quite a few came up with inclusions which could relate to archaeological finds or geoarchaeological information. The geotechnical case studies surrounding the site in question all showed signs of archaeological presence. The mentions of archaeology and how it aided this research project supported under the suggestion that borehole research can be done as a collaborative effort between the disciplines. The finds from the geotechnical case studies can be seen in Table 2 – Geotechnical Finds.

Table 2. Geotechnical Borehole Log.

Case	Depth	Organic	Shell	Flint	Pottery	Other
TW533410-179110	c. 18m	N/A	Oyster	Flints, No date	N/A	N/A
TW533530-178790	3.20m	Peat, Wood & Plant Material	N/A	N/A	N/A	N/A
TW533070 179020	2m	Large Wood pocked Waterlogged	N/A	N/A	Tile fragments	Brick Fragments
TW533700-179250	27.4m	N/A	Oyster	Green Flints	N/A	N/A
TW533060-179010	1.2m	Organic fragments	N/A	N/A	N/A	"Relics"
TW533390-179120	9,9m	N/A	Oyster	N/A	N/A	N/A
TW533600-178930	1.60m	Bones, organic clay	N/A	N/A	N/A	N/A
TW533460-178860	9.50m	Black carbon-aceous material	Oyster	N/A	Medieval	Pipe stems, ceramic drains
TW533360-179000	0.20m	Lumps of Dark brown Peat	N/A	N/A	N/A	CBM

Based on the case studies the potential loss of archaeology included:

5. Prehistoric landscape information
6. Dating information
7. Organic paleo-environmental proxy indicators and analysis of shells as a dietary means
8. Flint analysis, which could further contribute to the research on Neolithic use of the area

In summary:

9. 6 of 9 case studies had organic finds
10. 4 out of 9 case studies organic finds were related to landscape
11. 2 out of 9 case studies had Neolithic finds
12. 9 out of 9 case studies had archaeological material

The geotechnical case studies are interpretations from the geotechnical borehole scan logs, and as the geotechnical engineers behind the logs are not specialized in the field of archaeology and geoarchaeology, and have no formalized way of dealing with archaeological inclusions in their cores, leading to descriptions such as "relic". Credit goes to the geotechnical engineers for adding these descriptions on the record for researchers to interpret, as it is not currently a requirement in the British Standards. From the case studies the organic material found is high, which can be seen as a contributing argument to the prehistoric landscape surrounding the Case Study, especially seeing the ancient wood remains in a waterlogged environment found in the vicinity of the site (*TW533070 179020*) and the peat mentioned in several case studies (*TW533070 179020*, and *TW533360-179000*).

From the geoarchaeological and geotechnical case study findings, it can be concluded that it is high probability that the Case Study site in Bermondsey would contain organic material, with a high probability to relate to prehistoric landscape information, or paleo-environmental proxy indicators, as well as other organic artefacts. The surrounding case studies show that there is a high probability that some archaeological material would be contained in the Case Study, and that the lack of geoarchaeological boreholes performed on site resulted in a loss of potentially significant geoarchaeological information for the understanding of the prehistoric landscape and settlements.

5 DISCUSSION

A) Summary of Findings:

   i. Prehistoric Landscape and Settlements

      As found in both the archaeological and geotechnical data-sets, the prehistoric landscape can be found partially or intact in several surrounding locations of the Case Study. Sites such as Alaska Works (*MLO23838/ELO2701*) to the north-east indicates that there is potential for finds and deposits from this period to survive un-truncated on the gravels of the Case Study site, which is supported by archaeological case studies *MLO63702* and *MBZ06* seen in Table 1, and geotechnical case studies *TW533070 179020*, and *TW533360-179000* seen in Table 2. In addition to the prehistoric landscape, finds dating from the Neolithic settlements, as seen at Bricklayers Arms Goods Depot (*MLO63702*), to Roman farm settlements (*MLO105026*) can be found surrounding the Case Study. All of the case studies had archaeological material surviving, which provides a good estimate for the potential for geoarchaeological finds at the Case Study.

   ii. Using Surrounding Data to Predict Archaeology

      The research has shown the full potential of predicting the geoarchaeology on a site, by its extraction of information from nearby sites. The estimated finds were significantly organic, with wider implications of the ancient landscape of Bermondsey and settlement groups dating back to the Neolithic. There is no way to confirm the estimated losses without doing boreholes in the area discussed. Hopefully, the potential data estimation for the area can help fill in the knowledge gap as more archaeological work is done in Bermondsey, and a clearer image of the ancient landscape and its uses and changes appears.

   iii. Potential for Geotechnical Logs and Archaeology

      An interesting aspects of the research has been the discovery of archaeological mentions on geotechnical borehole scan logs, which show that there is potential to combine geoarchaeological and geotechnical research on development sites. The research found that there is already some mentions of archaeological "inclusions" from geotechnical engineers, but with no formalized language to describe their findings.

B) Meaning, Implications & Value of Findings

   i. Prehistoric Landscape and Settlements

      As shown in Tables 1 & 2 on numerous case studies, the area is of importance for its historic landscape and the information it could provide researchers with to understand the historical area of Bermondsey. By its implication, the case studies show that there is reason to suggest that parts of prehistoric landscapes or paleo-environmental proxy indicators could be found on the Case Study site, meaning that the decision to not do boreholes at the site lost valuable historical environmental information from the area. The implication from this finding shows the value of full archaeological investigations imbedded in a development project, and the potential loss incurred if proper archaeological research is not permitted on a site.

   ii. Using Surrounding Data to Predict Archaeology

      In terms of predicting geoarchaeological presence on a site, there are no guarantees. The only way to fully know what a site contains is by excavation and borehole research by commercial archaeologists teams. The research did find several indicators for what potential finds could occur on the site, based on surrounding finds and their preservation. In cases where there is no other archaeological information to gather, it is a way to ensure some knowledge of an area goes on record and contributes to the understanding of an area. The implication and value of this finding is to show that here are other ways to gather archaeological data on a site that could be considered of high importance, so that the site can be seen in relation to the area as a whole to decide land use and change over time.

iii. Potential for Geotechnical Logs and Archaeology
The implication of geotechnical engineers already attempting to describe archaeological additions in their boreholes signify a willingness and opportunity for collaborative efforts in the area with archaeologists. The implementation of synergistic collaborative efforts using boreholes as a medium between geoarchaeologists and geotechnical engineers should be considered as a large-scale change initiative with the strengths, weaknesses, opportunities and threats to be evaluated. The implications for such a collaborative effort would call for reclassification of tools and recording practices, clear guidelines, as well as disciplinary training. The potential value of this finding and collaborations could be enormous for development and research projects, in terms of time and cost savings, as well as creating a more wholesome geological framework of the area.

6 CONCLUSIONS

The research has found several areas of potential finds in the Case Study and identified the potential loss experienced on the site. For numerous reasons, the archaeological investigation of the site was not done to its full extent, and the archaeological research suffered for it. The research was based on the relevant geotechnical borehole logs and research done in the area, which was compared to archaeological finds in the same proximity to the Case Study. A total of 167 geotechnical borehole logs in the area were considered, and a total of 128 archaeological logs. From these logs 9 case studies from archaeology and 9 case studies from geotechnical engineering were chosen. Based on these case studies, the potential loss of archaeology included:

13. Prehistoric landscape information
14. Information regarding land use
15. Dating information
16. Flint analysis
17. Organic artefacts and other organic material surviving in the wet environment
18. Dating information helping to decide settlement movement and landscape change

In addition to the findings about the lost archaeology, the research has rightly pointed out that there should, and could, be more collaboration between geotechnical engineers and geoarchaeologists. Both disciplines perform analysis which are knowledge-intensive tasks (Stein, 1986:505), and as found by Goldberg and Machphail (2006:29) geoarchaeology, like geology and geotechnical engineering, is a field-based endeavor that relies on empirical data accumulated from a site (Caputo, 2003:688). In addition to this, the main source of information lost was from boreholes, which both disciplines undertake in a highly similar manner. In this Case Study, only the geotechnical borehole logs were permitted. If there had been some collaboration, or language adapted to take archaeological objects into account and common descriptions, the archaeologist could have gained some information purely from the borehole logs, or even better – the boreholes could have been done in a collaborative manner leaving both the geotechnical engineers and archaeologists with the results they needed.

The research finds that questions one can be answered through surrounding case studies, and identifies organic material as the significant loss from the site. The study has answered this question by the close analysis of borehole scan logs and archaeological fieldwork logs in the Bermondsey area, identifying the most relevant case studies and how it can be related to the Case Study in a coherent manner. In addition to this, the research has shown that there is an opportunity for collaborative work practices in borehole research between geotechnical engineers and geoarchaeologists, with the development of synergistic interdisciplinary work practices. This case study hopes to show the importance of full archaeological investigations on development sites, what loss limited archaeological research can bring, and crucially, to show how more collaborative work practices between geotechnical engineers and geoarchaeologists can result in more comprehensive research reports, by synergistic research practices which enables both disciplines to do the research required and enhances understanding. A number of barriers were identified for the interdisciplinary collaboration to work. However, the synergistic use of borehole research was identified as a knowledge integration tool providing wider reach and thus increased access to a wealth of knowledge, experience and resources throughout a development site. Interdisciplinary research can, thus, provide various time and cost saving at both project and organizational levels, and needs to be further researched.

To conclude, it is possible to predict potential archaeological finds and geoarchaeological information on a development site, which can help understand an area as a whole in a historic setting. An important factor for ensuring that the proper archaeological information is gathered from a site is to incorporate it with geotechnical investigations, which would enhance understanding and preservation and encourage interdisciplinary understanding, as well as save time and cost for projects and research.

REFERENCES

Caputo, V. «The Role of Geotechnical Engineering in the Preservation of our Architectural Heritage.» *Transactions on the Built Environment Vol. 66*, 2003.

City of London. «The Impact of Archaeology on Property Development.» *City of London Publications*. 2012. http://www.cityoflondon.gov.uk/business/economic-research-and-information/research-publications/Documents/2007-2000/The-Impact-of-Archaeology-on-Property-Development.pdf (funnet April 2018).

Cotton, J. Green, A. «Further Prehistoric finds from Greater London.» *London and Middlesex Archaeological Society, 55*, 2004.

D'Agostino, S. Tocco, G. «Archaeology and Geotechnical Engineering». I *Geotechnical Engineering for the Presrevation of Monuments and Historic sites*, av E. Flora, A. Lirer, S. Viggiani, C. Bilotta. London: Taylor & Francis Group, 2013.

Eisenhardt, K.M. Graebner, M.E. «Theory Building from Cases: Oppertunities and Challenges.» *The Academy of Management Journal* vol 50:1, 2007: 25–32.

Faraj, S. Xiao, Y. «Coordination in Fast-Response Organizations.» *Management Science vol 52:8*, 2006: 1155–1169.

Glaser, B. G. Straus, A. L. *Discovery of Grounded Theory: Strategies for Qualitative Reserach*. New York: Aldine de Gruyter, 1967.

Historic England. *National Infrastructure Development and Historic Environment Skills and Capacity 2015–33: An Assessment*. London, 2016.

Jørgensen, B. Emmitt, S. «Investigating the Integration of Deisgn and Construction from a "Lean" Perspective.» *Construction Innovation vol 9:2*, 2009: 225–240.

Knight, H. *Aspects of Medieval and later Southwark: Archaeological Excavations (1991–8) for the London Underground Limited Jibilee Line Extension Project*. MoLAS Monograph 13, MOLA, 2002.

Kruse, F. *Geotechnical and Geoenvironmental Site Investigations. Guide 11.*. London: BAJR Practical Guide Series, 2006.

Mackinder, A. *A Roman-British cemetery on Watling Street, excavations at 156 Great Dover Street, Southwark*. MOLA: London MoLAS Archaeology study series 4, 2000.

Ridgeway, V. «Prehistoric finds at Hopton street in Southwark.» *London Archaeologist 9:3*, 1999: 72–76.

Rogers, W. «Mesolithic and Neolithic flint tool-manufacturing areas buried beneath Roman Watling Street in Southwark.» *London Archaeologist 6:9*, 1990: 227–231.

Sidell, E.J. Wilkinson, K. Scaife, R. Cameron, N. *The Holocene Evolution of the London Thames.* MOLA: MoLAS Monograph 5, 2000.

Stein, J.C. «Coring Archaeological Sites.» *American Antiquity Vol. 51*, 1986: 505–527.

Viggiani, C. «Cultural Heritage and Geotechnical Engineering». I *Geotechnical Engineering for the Preservation of Monuments and Historic Sites*, av E. Flora, A. Lirer, S. Viggiani, C. Bilotta. London: Taylor & Francis Group, 2013.

Tunnels and Underground Cities: Engineering and Innovation meet Archaeology, Architecture and Art, Volume 1: Archeology, Architecture and Art in underground construction – Peila, Viggiani & Celestino (Eds)
© 2020 Taylor & Francis Group, London, ISBN 978-0-367-46574-2

Author Index

Anagnostou, G. 147
Arigoni, A. 127
Avanoğlu-Çetin, B. 3
Avataneo, G. 199

Bellone, A. 188
Bortolussi, A. 42
Brenci, F. 12

Callari, C. 22
Cassani, G. 108
Cavuoto, F. 32
Colombo, M.N. 42
Corbo, A. 32

Dati, G. 199
Diederichs, M. 52
D'Angelo, M. 167, 178

Ferreira, P. 209
Foti, V. 167
Foufas, P. 147
Frandi, F. 62

Gatti, M. 108
Gazzola, S. 108
Grossauer, K. 127

Hashemi, S. 72, 80
Hohermuth, M. 127
Hutchinson, D.J. 52

Iannotta, F. 118

Kutkan-Öztürk, Y. 90

Laudato, M. 98
Lunardi, G. 108

Manassero, V. 32
Manfredi, E. 118
Modetta, F. 127

Noce, E. 42

Öztürk, Ö. 3

Pedemonte, S. 137
Piangatelli, M.A. 188
Pizzarotti, E.M. 137

Ricci, M. 199
Rizos, D. 147
Romagnoli, G. 157
Romani, E. 167, 178
Romano, F. 118
Rossano, F. 188
Russo, G. 32

Saviani, S. 127
Sorge, R. 178
Sully, D. 209

Vanfiori, S. 118
Vassilakopoulou, G. 147
Virano, M. 199
Vonstad, F.K. 209